Practical Applications Circuits Handbook

Practical Applications Circuits Handbook

Anne Fischer Lent
Peterborough, New Hampshire

Stan Miastkowski
Peterborough, New Hampshire

ACADEMIC PRESS, INC.
Harcourt Brace Jovanovich, Publishers
San Diego   New York   Berkeley   Boston   London   Sydney   Tokyo   Toronto

Copyright © 1989 by Academic Press, Inc.
All Rights Reserved.
No part of this publication may be reproduced or transmitted in any form or by any means, electronic or mechanical, including photocopy, recording, or any information storage and retrieval system, without permission in writing from the publisher.

Academic Press, Inc.
San Diego, California 92101

United Kingdom Edition published by
Academic Press Limited
24–28 Oval Road, London NW1 7DX

Library of Congress Cataloging-in-Publication Data

Lent, Anne Fischer.
    Practical applications circuits handbook / Anne Fischer Lent and Stan Miastkowski.
      p. cm.
    Includes index.
    ISBN 0-12-443775-3 (alk. paper)
    1. Electronic circuits--Handbooks, manuals, etc.   I. Miastkowski, Stan. II. Title.
TK7867.L47   1989
621.381'32--dc20   89-32344
                                                      CIP

Printed in the United States of America
89 90 91 92   9 8 7 6 5 4 3 2 1

This book is dedicated to Greg, Steven, Dexter, Kathe, and Pepper, who endured numerous solitary evenings and weekends in the name of electronic literacy.

Contents

Preface	xix
Acknowledgments	xxi
Abbreviations	xxiii

Chapter 1
Audio Amplifiers

310-milliwatt Transformerless Audio Amplifier	2
Basic Class B Audio Amplifier	2
3- to 5-watt Audio Amplifiers	3
12-watt Audio Amplifier	5
15-, 20-, and 25-watt Darlington Amplifiers	6
20-watt Audio Amplifier	7
7- to 35-watt Audio Amplifiers	8
High Power Audio Amplifier	11
100-watt Audio Amplifier	13
Gain-Controlled Stereo Amplifier	13
Ceramic Pickup Amplifier	14

Chapter 2
Amplifiers

Variable Gain Amplifier	16
Lock-In Amplifier	17
DC-Coupled Inverting Amplifier	18
DC-Coupled Noninverting Amplifier	19
Chopper-Stabilized Instrumentation Amplifier	20
+ 5-volt Precision Instrumentation Amplifier	21
Precision High-Speed Op Amp	22
Large Signal-Swing Output Amplifier	23
Ultra-Precision Instrumentation Amplifier	23
Precision Isolation Amplifier	24
DC-Stabilized Low Noise Amplifier	26
AC-Coupled Inverting Amplifier	27

Current Mode Feedback Amplifier	28
Fast DC-Stabilized FET Amplifier	29
Voltage-Controlled Amplifier	30
Fast-Stabilized Noninverting Amplifier	31
Gain-Trimmable Wideband FET Amplifier	32
Low-Power Voltage-Boosted Output Op Amp	33
Single-Supply Differential Bridge Amplifier	34
Three-Channel Separate-Gain Amplifier	34
N-Stage Parallel-Input Amplifier	35
Precision Amplifier with Notch	36
Differential Input/Output Amplifier	36

Chapter 3
Audio Circuits

Microphone Preamplifier	38
Music Synthesizer	39
Voltage-Controlled Attenuator	40
Automatic Level Control	41
Tape Head Preamplifier	41
Variable-Slope Compressor–Expander	42
Hi-Fi Compandor	43
RIAA/NAB Compensation Preamplifier	44
Four-Input Stereo Source Selector	45
RIAA-Equalized Stereo Preamplifier	46
Fast Attack, Slow Release Hard Limiter	47
Audio Decibel Level Detector	48
Single-Frequency Audio Generator	49
AGC Amplifier	49
Audio Equalizer	50
Rumble/Scratch Filter	52
RIAA Preamplifier	53
Stereo Volume Control	54

Chapter 4
Automotive Circuits

Tachometer	56
Auto Burglar Alarm	57
Speed Warning Device	58
Automobile Voltage Regulator	59
AM/FM Car Radio	60
AM Car Radio	61
6-watt Audio Amplifier	62

Chapter 5
Battery Circuits

6-volt Battery Charger	64
12-volt Battery Charger	64
Wind-Powered Battery Charger	65
Battery Backup Regulator	66
Battery Splitter	66
High-Current Battery Splitter	67
Battery Cell Monitor	67
Micropower Switching Regulator	68
Sine Wave Output Converter	69
Voltage Doubler	70
Low-Power Flyback Regulator	70
Battery Monitor for High-Voltage Charging Circuits	71
Thermally Controlled NiCad Battery Charger	71
Single-Cell Up Converter	72
Low Dropout 5-volt Regulator	73
Computer Battery Control	75

Chapter 6
Control Circuits

Zero-Voltage, On–Off Controller with Isolated Sensor	77
"Tachless" Motor Speed Controller	78
Wall-Type Thermostat	79
Low-Cost Touchtone Decoder	80
On–Off Touch Switch	81
Precision Temperature Controller	82
Alarm System	83
Overload-Protected Motor Speed Controller	83
Fan-Based Temperature Controller	84
Line-Isolated Temperature Controller	85
Dual-Output Over–Under Temperature Controller	86
Dual Time-Constant Tone Decoder	86
3-Phase Sine Wave Circuit	87
Sensitive Temperature Controller	88
Temperature Control	89
Squelch Control	90
Frequency Meter with Low-Cost Lamp Readout	91
Wide-Band Tone Detector	92
Motor Speed Control	93
Automatic Light Control	93

Chapter 7
Converters

1-Hz–1.25-MHz Voltage-to-Frequency Converter	95
1-Hz–30-MHz Voltage-to-Frequency Converter	96
16-bit Analog-to-Digital Converter	97
Cyclic Analog-to-Digital Converter	98
Current-to-Voltage Converter	98
Offset Stabilized Voltage-to-Frequency Converter	99
Current Loop Receiver/Transmitter	100
Thermocouple-to-Frequency Converter	101
9-bit Digital-to-Analog Converter	102
TTL-to-MOS Converter	102
Temperature-to-Frequency Converter	103
10-Bit 100µA Analog-to-Digital Converter	104
Micropower 12-Bit 300-µsec Analog-to-Digital Converter	105
Fully Isolated 10-Bit A/D Converter	106
V/F-F/V Data Transmission Circuit	107
Quartz-Stabilized Voltage-to-Frequency Converter	107
2.5-MHz Fast-Response V/F Converter	108
Push–Pull Transformer-Coupled Converter	109
Ultra-High Speed 1-Hz to 100-MHz V/F Converter	110
Low-Power 10-KHz V/F Converter	111
ECL-to-TTL and TTL-to-ECL Converters	112
Low-Power 10-Bit A/D Converter	113
Basic 12-Bit 12-µsec Successive Approximation A/D Converter	114
Wide-Range Precision PLL Frequency-to-Voltage Converter	115
Centigrade-to-Frequency Converter	116
Micropower 10-kHz V/F Converter	118
Micropower 1-MHz V/F Converter	119
RMS-to-DC Converter	120

Chapter 8
Filter Circuits

Low-Pass Filter	122
Loop Filter	122
Voltage-Controlled, Second-Order Filter	123

CONTENTS   xi

Voltage-Controlled Filters	124
Tracking Filter	125
Digitally Tuned Switched-Capacitor Filter	126
10-Hz, Fourth-Order Butterworth Low-Pass Filter	127
Clock-Tunable Notch Filter	127
Bi-Quad Filter	128
Voltage-Controlled Circuits	129
Multi-Cutoff-Frequency Filter	130
Lowpass Filter with 60-Hz Notch	131
60-Hz Reject Filter	131
Voltage-Controlled Filter	132

Chapter 9
Function Generators

Function Generator	134
Logic Function Generator	135
Triangle Wave Oscillator	136
Triangle/Square-Wave Generator	136
Single-Chip Function Generator	137
Triangle-to-Sine Wave Converter	138
Waveform Generator	139
Function Generator	140

Chapter 10
Measurement Circuits

Linear Thermometer	142
Accelerometer	143
Sequential Timer	144
Thermocouple Amplifier	145
Narrow-Band Tone Detector	146
Negative Current Monitor	147
Acoustic Thermometer	148
Transmitting Thermometer	150
Magnetic Tachometer	150
Linear Thermometer	151
Voltmeter	152
Low-Flow-Rate Thermal Flowmeter	153
Presettable Timer with Linear Readout	154
Thermally Based Anemometer	155

Relative Humidity Signal Conditioners	156
Digital Thermometer	157
Sine-Wave Averaging AC Current Monitor	158
Level Transducer Digitizer	159
Thermometer Circuit	160
Four-Channel Temperature Sensor	161
Photodiode Digitizer	162
Field Strength Meter	164
Tachometer	164
Long-Duration Timer	165
Digital Thermometer	165

Chapter 11
Microprocessor Circuits

8-Bit Serial-to-Parallel Converter	167
Voltage-Sag Detector	167
Power-Loss Detection Circuit	168
CRT Driver	169
Clock Regenerator	170
Bidirectional Bus Interface	171
Cheapernet/Ethernet Interface	172

Chapter 12
Miscellaneous Circuits

Intercom	174
Buffered Output Line Driver	175
Low-Voltage Lamp Flasher	175
Digital Clock with Alarm	176
Astable Single-Supply Multivibrator	177
Frequency Synthesizer	177
Fed Forward, Wideband DC-Stabilized Buffer	178
Three Clock Sources	179
Freezer Alarm	180
Fiber Optic Receiver	181
Micropower Sample-Hold	182
Frequency Output Analog Divider	183
Protected High Current Lamp Driver	183
Tone Transceiver	184
Variable Shift Register	185
Fast, Precision Sample-Hold Circuit	186
Precision PLL	187
Two Peak Circuits	188

Chapter 13
Modem Circuits

Bell 212A Modem	190
300-bps Full-Duplex Modem	192
1200-bps Modem	193
Full-Duplex 300/1200-bps Modem System	194
2400-bps CCITT Modem	196
Auto Dialer Modem	198
2400-bps Stand-Alone Intelligent Modem	199
Notch Filters	200
Full-Duplex FSK Modem	201
High-Speed FSK Modem	202
Power Line Modem	204

Chapter 14
Optoelectronics Circuits

LED Driver	206
Photodiode Amplifiers	207
Light Amplifier	207
Infrared Remote-Control System	208
Sensitive Photodiode Amplifier	209
Photo Diode Detector	210
High-Speed Photodetector	210
Balanced Pyroelectric Infrared Detector	211
100-dB Range Logarithmic Photodiode Amplifier	212

Chapter 15
Oscillator Circuits

1- to 10-MHz and 1- to 25-MHz Crystal Oscillators	215
Crystal-Stabilized Relaxation Oscillator	216
L-C Tuned Oscillator	217
Temperature-Compensated Crystal Oscillator	218
Stable RC Oscillator	219
Reset Stabilized Oscillator	220
High-Current Oscillator	220
Crystal-Controlled Oscillator	221
Synchronized Oscillator	221
Voltage-Controlled Crystal Oscillator	222
Voltage-Controlled Oscillator	223

1-Hz to 1-Mhz Sine Wave Output Voltage-Controlled Oscillator	224
Low-Power Temperature Compensated Crystal Oscillator	225
Digitally Programmable PLL	226
First-Harmonic (Fundamental) Oscillator	227
A Low-Frequency Precision RC Oscillator	227
Low-Distortion Sinewave Oscillator	228
Wein Bridge Oscillator	229
Temperature-Compensated Crystal Oscillator	230

Chapter 16
Power Supply Circuits

Basic Power Supply	232
TTL Power Supply Monitor	233
Programmable Voltage/Current Source	234
Regulated Negative Voltage Converter	235
Current Monitor	235
Two Rectifiers	236
Bridge Amplifier Load Current Monitor	237
High-Efficiency Rectifier Circuit	238
Current-Limited 1-amp Regulator	238
5-V Regulator	239
5-V Regulator with Shutdown	239
Dual Output Regulator	240
Regulator with Output Voltage Monitor	241
RMS Voltage Regulator	242
Switching Regulator	243
Micropower Post-Regulated Switching Regulator	243
High-Current Switching Regulator	244
Switching Preregulated Linear Regulator	245
Low-Power Switching Regulator	246
Fully Isolated −48-V to 5-V Regulator	247
Inductorless Switching Regulator	248
Single Inductor, Dual-Polarity Regulator	249
Dual Tracking Voltage Regulator	250
Current Regulator	251
Negative-Voltage Regulator	252
High-Voltage Regulator	253
Switching Regulator	254
High-Temperature +15-V Voltage Regulator	255
7.5-A Variable Regulator	255
Adjustable Regulator	256
High-Efficiency Regulator	257

Chapter 17
Receiving Circuits

SCA Demodulator	259
FM Tuner	260
Narrow-Bandwidth FM Demodulator	261
Clock Regenerator	262
10.8-MHz FSK Decoder	263
Narrow-Band FM Demodulator	264
Linear FM Detector	264
10.7-MHz IF Amplifier	265
88- to 108-MHz FM Front End	266
Balanced Mixer	267
4–20-mA Current Receiver	267

Chapter 18
Signal Circuits

Traffic Flasher	269
Tone Burst Generator	270
2.2-watt Incandescent Lamp Driver	270
Wide-Range Automatic Gain Control	271
Low-Droop Positive Peak Detector	271
High-Speed Peak Detector	272
Single-Burst Tone Generator	273
Bandpass Filter for a Multi-Channel Tone Detector	274

Chapter 19
Telephone Circuits

Line-Powered Tone Ringer	277
Tone Telephone	278
Basic Telephone Set	279
Nonisolated 48- to 5-V Regulator	280
Programmable Multitone Telephone Ringer	281
Featurephone with Memory	282
Audio Frequency Sweepers	284
Bell System 202 Date Encoder and Decoder	286
Ring Signal Counter	287
Ring Detector Circuits	288

Chapter 20
Test Circuits

0–5 Amp, 7–30 Volt Laboratory Supply	291
Pulse Generator	292
Wide-Range Frequency Synthesizer	293
Analog Multiplier with 0.01% Accuracy	294
Leakage Current Monitor	295
Computer-Controlled Digitizer	296
Digitally Programmable Waveform Generator	298
Voltage Reference	298
1-nsec Rise Time Pulse Generator	299
50-MHz Trigger	299
Active and Passive Signal Combiners	300
Voltage References	301
Precision Programmable Voltage Reference	302
Cable Tester	303
Low-Frequency Pulse Generator	304
12-bit Digitally Programmable Frequency Source	305
Voltage Programmable Pulse Generator	306
Ultraprecision Variable Voltage Reference	307
Low-Noise Instrumentation Amplifier	307
Gain-Ranging Amplifier	308

Chapter 21
Transmitting Circuits

80-watt, 175-MHz Transmitter	310
Balanced Modulator	312
Frequency Doublers	313
Amplitude Modulator	314
Broadband 160-watt Linear Amplifier	315
Three VHF Amplifiers	316
4- to 20-mA Current Loop Transmitter	318

Chapter 22
Video Circuits

Analog Multiplier Video Switch	321
Video Amplifier	322
Color Video Amplifier	323
Wide Bandwidth VCA for Video	324
Video Line Driving Amplifier	325

NTSC Video/Data Inlay Chip	326
Broadband Video Amplifier	327
IF Amplifier and Detector	328
Wide-Band AGC Video Amplifier	330
NTSC Color Decoder	331
Wide-Band Video Amplifier	332
PAL/NTSC Decoder	333

Chapter 23
Voltage Circuits

Fast, Synchronous Rectifier-Based AC/DC Converter	335
Visible Voltage Indicator	336
12-nsec Circuit Breaker	336
Offset Stabilized Comparator	337
−5-volt Bus Monitor	338
+6- to +15-volt Converter	339
Regulated Voltage Up Converter	340
Precision Voltage Inverter	340
50-MHz Bandwidth RMS-to-DC Converter	341
Standard-Grade Variable Voltage Reference	342
Dual-Voltage Tracking Regulator	342
Isolated Power Line Monitor	343
Dual-Limit Threshold Detector	343
LVDT Signal Conditioner	344

Index	345

Preface

Whether you're a professional engineer, a student, or a hobbyist, you've undoubtedly found that getting practical information on the types of circuits you need to use in common day-to-day applications is difficult. Textbooks and technical books tend toward the theoretical rather than the practical. Component manufacturers do provide a wide range of practical applications data, but there are scores of manufacturers and literally thousands of different application notes, data books, and reference guides. The time required to gather all this material (not to mention the money you'd spend) makes gathering extensive applications data impractical for most people. In addition, you'd need an extra office just to store it all.

In this book, we've done the leg work for you. The *Practical Applications Circuits Handbook* contains some 350 circuit designs and applications culled from manufacturer data. The purpose is to avoid having to "reinvent the wheel" when designing an application. You can use most of the circuits as they are, since they've ostensibly been tested by the manufacturers to make the best use of their particular components. You can also use the circuits as a jumping-off point for additional design since many can be modified by changing a few components or component values.

The circuits in this book are largely taken from the most up-to-date manufacturer literature available. There are many examples using components that are at the cutting edge of chip design. But we've also included some of the "oldies but goodies," time-proven circuit designs that are still the best practical examples in this age of high-tech. Also, don't be surprised to see some repetition in circuit types. In many cases, different manufacturers suggest different designs for cir cuits that are intended to perform the same function. Comparing these designs can be a valuable lesson in varying approaches to circuit design.

You'll undoubtedly notice that there's a large difference in the design of the schematics in this book. That's because we've reproduced the circuit diagrams from their original source, making sure that nothing has been lost in the translation. Where possible, we've also included the original manufacturer parts list and the significant design equations that will help you understand how the circuit works or allow you to modify the circuit to your particular needs should you wish to do so.

Each circuit includes a brief description of its overall design, and (where necessary) background information on concepts and components. But make no mistake, this is no theoretical tome that's destined to gather dust on the shelf. As this book's title implies, this is a *practical* guide to application circuits. It's designed to allow you to quickly find a solution to a particular design problem. If you want or need more-detailed data, there are plenty of books available that delve deeply into circuit theory.

We've also included with each circuit a citation of its original source. Because of space limitations, detailed design information for some of the circuits had to be omitted. To obtain these data (or if you just want to learn more about a particular application), consult the original source. You'll find the addresses of the sources given in the Acknowledgments section (p. xxi). In many cases, manufacturers will be more than happy to send you a reasonable number of their application notes or data books free of charge, especially if they're requested on company letterhead. Some manufacturers charge for design data, and others offer a subscription service that will ensure you always receive the latest detailed design data.

We hope that you'll find this book useful. It should save you hours or even days of searching when you need to find a circuit for your specific application.

Anne Fischer Lent
Stan Miastkowski

Acknowledgments

The publication of this book was made possible because of the cooperation of the following companies. If you wish to obtain the original sources for the circuits shown in this book, you should contact them directly. The authors express their gratitude to the following:

Advanced Micro Devices, Inc., 901 Thompson Place, P.O. Box 3453, Sunnyvale, California 94088

Analog Devices, Inc., Two Technology Way, Norwood, Massachusetts 02062

Burr-Brown Corporation, International Airport Industrial Park, P.O. Box 11400, Tucson, Arizona 85734

EXAR Corporation, 2222 Qume Drive, San Jose, California 95131

Linear Technology Corporation, 1630 McCarthy Boulevard, Milpitas, California 95035

Motorola Semiconductor Products, Inc., P.O. Box 20912, Phoenix, Arizona 85036

NEC Electronics, 401 Ellis Street, P.O. Box 7241, Mountain View, California 94039

RCA Solid State Division, P.O. Box 3200, Somerville, New Jersey 08876

Signetics Corporation, 811 E. Arques Ave., P.O. Box 3409, Sunnyvale, California 94088-3409

All circuits in this book have been reproduced with the express written permission of each of the companies listed above. Any additional reproduction of any circuit without the express written permission of company involved is prohibited.

All the above companies reserve all rights to the circuits in this book. Any inquiries regarding reproduction or other use of the circuits should be addressed directly to the company involved.

Abbreviations

The following abbreviations are used in the schematics and circuit descriptions that you'll find in this book. In some cases, you'll notice some inconsistencies in the style of the abbreviations. This is because the schematics have been compiled from a variety of different sources.

A	ampere	F/V	frequency/voltage
AC	alternating current	FSK	frequency-shift keying
AC/DC	AC or DC operation	GHz	gigahertz
A/D	analog digital	H	henry
AFC	automatic frequency control	Hz	hertz
AGC	automatic gain control	IC	integrated circuit
AM	amplitude modulation	IF	intermediate frequency
ASCII	American Standard Code for Information Interchange	IMD	intermodulation distortion
		I/O	input/output
ASK	amplitude-shift keying	K	kilohm or kilobyte (1024 bytes)
BCD	binary-coded decimal	kHz	kilohertz
bps	bits per second	kw	kilowatt
C	centigrade or capacitor	L	inductor
CCO	current-controlled oscillator	L-C	inductor-capacitor
cm	centimeter	LED	light-emitting diode
CMOS	complementary metal-oxide semiconductor	LSB	least significant bit
		LSI	large-scale integration
CRT	cathode ray tube	LVDT	linear-variable differential transformer
CT	center-tapped		
D	diode	m	milli
D/A	digital-to-analog	M	megohm or megabyte
DAC	digital-to-analog converter	mA	milliampere
dB	decibel(s)	meg	megohm or megabyte
DC	direct current	MSB	most significant bit
D/F	digital/frequency	MHz	megahertz
DIFET	dielectrically isolated field-effect transistor	mL	milliliter
		mm	millimeter
DIP	dual inline package	MOS	metal-oxide semiconductor
DPDT	double-pole, double-throw switch	ms	millisecond
DPST	double-pole, single-throw switch	mW	milliwatt
DTMF	dual-tone multi-frequency	ns	nanosecond
DVM	digital voltmeter	NTSC	national television standards committee
ECL	emitter-coupled logic		
EMF	electromotive force	pF	picofarad
EMI	electromagnetic interference	p-p	peak to peak
EPROM	erasable PROM	PC	printed circuit or personal computer
F	farenheit or farad		
FET	field-effect transistor	pot	potentiometer
FM	frequency modulation	PLL	phase-locked loop

ppm	parts per million	SPST	single-pole, single-throw switch
PSK	phase-shift encoding	T	transformer
PMOS	P-channel MOS	TMOS	enhanced special power FET
PROM	programmable read-only memory	TTL	transistor-transistor logic
PUT	programmable unijunction transistor	uA	microampere
		UART	universal asynchronous receiver-transmitter
Q	transistor or quality factor		
R	resistor	UHF	ultra-high frequency
RAM	random-access memory	UL	Underwriter's Laboratories
R/C	resistor/capacitor	uS	microsecond
rf	radio frequency	uV	microvolt
RFI	radio frequency interference	V	volt
RGB	red-green-blue	VAC	volts alternating current
RIAA	Recording Industry Association of America	VC	voltage-controlled
		VCA	voltage-controlled amplifier
RMS	root-mean square	VCO	voltage-controlled oscillator
ROM	read-only memory	VDC	volts direct current
RPM	rotations per minute	V/F	voltage/frequency
RTD	resistance temperature detector	VHF	very-high frequency
SCA	subsidiary-carrier authorization	VU	volume unit
SAR	successive approximation converter	W	watt
SCR	silicon-controlled rectifier	X	times
SPAT	single-pole, double-throw switch		

Chapter 1

Audio Amplifiers

310-milliwatt Transformerless Audio Amplifier
Basic Class B Audio Amplifier
3- to 5-watt Audio Amplifiers
12-watt Audio Amplifier
15-, 20-, and 25-watt Darlington Amplifiers
20-watt Audio Amplifier
7- to 35-watt Amplifiers
High Power Audio Amplifier
100-watt Audio Amplifier
Gain-Controlled Stereo Amplifier
Ceramic Pickup Amplifier

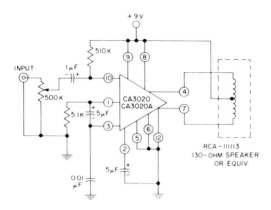

310-milliwatt Transformerless Audio Amplifier

You can use this circuit as a highly efficient class B audio power output circuit in such applications as communications systems, AM or FM radios, tape recorders, intercoms, or linear mixers. Because it's transformerless, it's particularly useful for portable applications. This amplifier gives 310 mV output for an input of 45mW and has a high impedance of 50,000 ohms. Source: W. M. Austin and H. M. Kleinman, "Application of the RCA CA3020 and CA3020A Integrated-Circuit Multi-Purpose Wide-Band Power Amplifiers," Application Note ICAN-5766, RCA Solid State.

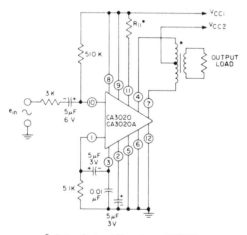

* Better Coil and Transformer DF108A, Thordarson TR-192, or equivalent.
• see text and tables.

Basic Class B Audio Amplifier

This simple amplifier circuit gives you a power output of either 1/2 or 1 watt, depending on whether you're using the RCA CA3020 or CA3020A. It can be used with a wide range of voltages. In this circuit, the emitter–follower stage at the input is used as a buffer amplifier to provide a high input impedance. The extended frequency range of the IC requires the high-frequency AC bypass capacitor used at the input. Otherwise, oscillation could occur at the stray resonant frequencies of the external components. Source: W. M. Austin and H. M. Kleinman, "Application of the RCA CA3020 and CA3020A Integrated-Circuit Multi-Purpose Wide-Band Power Amplifiers," Application Note ICAN-5766, RCA Solid State.

1. AUDIO AMPLIFIERS

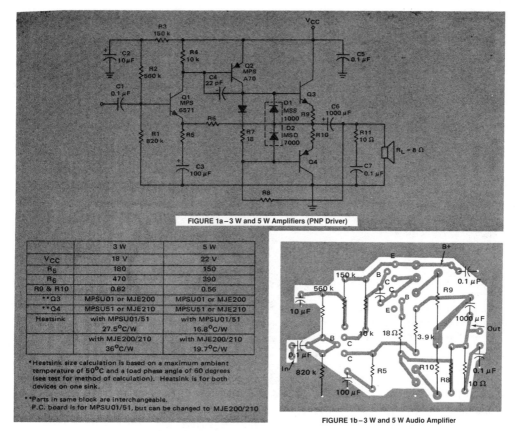

FIGURE 1a – 3 W and 5 W Amplifiers (PNP Driver)

	3 W	5 W
V_{CC}	18 V	22 V
R_5	180	150
R_6	470	390
R9 & R10	0.82	0.56
**Q3	MPSU01 or MJE200	MPSU01 or MJE200
**Q4	MPSU51 or MJE210	MPSU51 or MJE210
Heatsink	with MPSU01/51 27.5°C/W	with MPSU01/51 16.8°C/W
	with MJE200/210 36°C/W	with MJE200/210 19.7°C/W

*Heatsink size calculation is based on a maximum ambient temperature of 50°C and a load phase angle of 60 degrees (see test for method of calculation). Heatsink is for both devices on one sink.

**Parts in same block are interchangeable.
P.C. board is for MPSU01/51, but can be changed to MJE200/210

FIGURE 1b – 3 W and 5 W Audio Amplifier

TABLE 1a – Amplifier Performance Characteristics

Reference Figure 1	3 W 18 Vdc	5 W 22 Vdc
1. Idle Current (normal no-signal)	20 mA	50 mA
2. Current Drain at Rated Power	275 mA	360 mA
3. Typical Input Impedance	280 kohms	280 kohms
4. THD at Rated Output Power 20 Hz or kHz to 20 kHz	< 1%	< 1%
5. IM Distortion at 60 and 7000 Hz 4:1 ratio at Rated Power	< 1%	< 1%
6. -3 dB Bandwidth	20 Hz-290 kHz	20 Hz-325 kHz
7. Typical input sensitivity for rated output power	0.250 VRMS	0.250 VRMS
8. Maximum output power at 5% THD without current limiting	4.2 Watts	7.03 Watts
9. Maximum output power at 5% THD with current limiting	4.06 Watts	6.66 Watts
10. Power Supply Ripple Rejection	34 dB	32 dB
11. Short Circuit Power Supply Current with Current Limiting	750 mA	1 A

3- to 5-watt Audio Amplifiers

These two circuits show a general design for constructing 3- to 5-watt audio amplifiers using standard plastic-encapsulated transistors. The first design uses PNP transistors, the second NPN. These designs are only starting points, and no particular effort has been made to optimize the circuit performance. Source: "Basic Design of Medium Power Audio Amplifiers," Motorola Semiconductor Products, Application Note AN-484A, Motorola, Inc.

1. AUDIO AMPLIFIERS

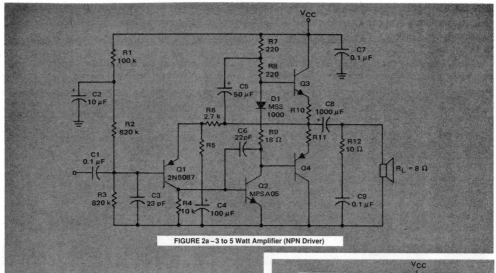

FIGURE 2a – 3 to 5 Watt Amplifier (NPN Driver)

Component	3 W	5 W
V_{CC}	17	22
R_5	120 Ω	100 Ω
R10 & R11	0.82	0.56
Q3*	MPSU01 / MJE200	MPSU01 / MJE200
Q4*	MPSU51 / MJE210	MPSU51 / MJE210
**Heatsink	MPSU01/51 27.5°C/W	MPSU01/51 16.8°C/W
**Heatsink	MJE200/210 36°C/W	MJE200/210 19.7°C/W

*Parts in same block are interchangeable. P.C. board is for MJE200/210, but can be easily changed to MPSU01/51.

**Heatsink size calculation is based on a maximum ambient temperature of 50°C and a load phase angle of 60 degrees (see test for method of calculation) Heat sink is for both devices on one sink.

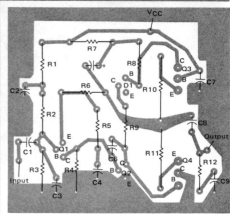

FIGURE 2b – P.C. Board 3 W and 5 W Amplifier (Copper Side)

TABLE 2a – Amplifier Performance Characteristics

Reference Figure 2	3 W 18 Vdc	5 W 22 Vdc
1. Idle Current (nominal no-signal)	20 mA	54 mA
2. Current Drain at Rated Power	285 mA	365 mA
3. Typical Input Impedance	300 kohms	320 kohms
4. THD at Rated Output Power 20 Hz or 1 kHz 20 kHz	< 1%	< 1%
5. IM Distortion at 60 and 7000 Hz 4:1 ratio at Rated Power	< 1%	< 1%
6. –3 dB Bandwidth	20 Hz-220 kHz	20 Hz-150 kHz
7. Typical input sensitivity for rated output power	0.22 VRMS	0.23 VRMS
8. Maximum output power at 5% THD without current limiting	4.10 Watts	6.8 Watts
9. Maximum output power at 5% THD with current limiting	4.06 Watts	6.65 Watts
10. Power Supply Ripple Rejection	24 dB	36.4 dB
11. Short Circuit Power Supply Current with Current Limiting	800 mA	1 Amp

1. AUDIO AMPLIFIERS

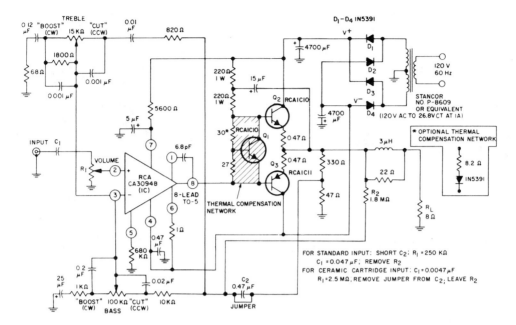

12-watt Audio Amplifier

This circuit can provide 12 watts of audio power output into an 8-ohm load with intermodulation distortion (IMD) of 0.2% when 60 Hz and 2 kHz signals are mixed in a 4 : 1 ratio. The large amount of loop gain and the flexibility of feedback arrangements with the RCA CA3094 monolithic programmable power switch/amplifier make it possible to incorporate the tone controls into a feedback network that's closed around the entire amplifier system. This results in an excellent signal-to-noise ratio. Hum and noise are typically 700 microvolts (83 dB down) at the output. Source: L. R. Campbell and H. A. Wittlinger, "Some Applications of a Programmable Power Switch/Amplifier," Application Note ICAN-6048, RCA Solid State.

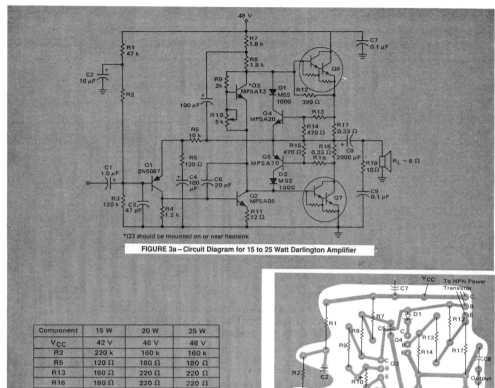

FIGURE 3a – Circuit Diagram for 15 to 25 Watt Darlington Amplifier

Component	15 W	20 W	25 W
V_{CC}	42 V	46 V	48 V
R2	220 k	160 k	160 k
R5	120 Ω	180 Ω	180 Ω
R13	180 Ω	220 Ω	220 Ω
R16	180 Ω	220 Ω	220 Ω
Q6	MJE800	TIP120	TIP120
Q7	MJE700	TIP125	TIP125
*Heatsink	6.5°C/W	4.8°C/W	3.6°C/W
R8	Adjusted for 20-30 mA Collector Current for Q6		

*Heatsink size calculated is based on a maximum ambient temperature of 50°C and a load phase angle of 60 degrees (see text for method of calculation). The heatsinks for both devices on one sink.

FIGURE 3b P.C. Board for 15 W, 20 W and 25 W Amplifier. (Copper Side)

TABLE 3a – Amplifier Performance Characteristics

Reference Figure 3	15 W 42 Vdc	20 W 46 Vdc	25 W 48 Vdc
1. Idle Current (nominal no-signal)	20 mA	20 mA	20 mA
2. Current Drain at Rated Power	640 mA	760 mA	800 mA
3. Typical Input Impedance	50 k	56 k	56 k
4. THD at Rated Output Power 20 Hz or 1 kHz to 20 kHz	< 1%	< 1%	< 1%
5. IM Distortion at 60 and 7000 Hz 4:1 ratio at Rated Power	< 1%	< 1%	< 1%
6. -3 dB Bandwidth	23 Hz-170 kHz	27 Hz-180 kHz	28 Hz-190 kHz
7. Typical input sensitivity for rated output power	200 mV	220 mV	230 mV
8. Maximum output power at 5% THD without current limiting	21.2 Watts	28.5 Watts	32.4 Watts
9. Maximum output power at 5% THD with current limiting	21 Watts	28 Watts	30 Watts
10. Power Supply Ripple Rejection	29 dB	34 dB	33 dB
11. Short Circuit Power Supply Current with Current Limiting	1.5 Amps	1.5 Amps	1.5 Amps

15-, 20-, and 25-watt Darlington Amplifiers

Here's a general design for an audio amplifier that outputs 15, 20, or 25 watts depending on the supply voltage and the choice of components. Source: "Basic Design of Medium Power Audio Amplifiers," Motorola Semiconductor Products, Application Note AN-484A, Motorola, Inc.

1. AUDIO AMPLIFIERS

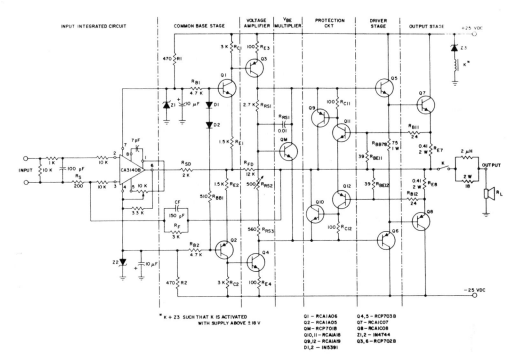

20-watt Audio Amplifier

Because they must closely reproduce the quality of the original input, hi-fi amplifiers have stringent design specifications. This 20-watt design illustrates just one of the many ways to design an amplifier for the best possible performance. See the original application note for a detailed discussion of overall design considerations for hi-fi amplifiers. Source: "A Practical Approach to Audio-Amplifier Design," Power Transistors, Application Note AN-6688, RCA Solid State.

1. AUDIO AMPLIFIERS

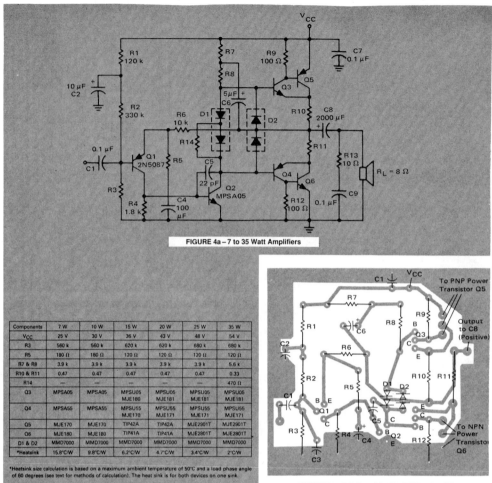

FIGURE 4a – 7 to 35 Watt Amplifiers

Components	7 W	10 W	15 W	20 W	25 W	35 W
V_{CC}	25 V	30 V	36 V	43 V	48 V	54 V
R3	560 k	560 k	620 k	620 k	680 k	680 k
R5	180 Ω	180 Ω	120 Ω	120 Ω	120 Ω	120 Ω
R7 & R8	3.9 k	3.9 k	3.9 k	3.9 k	3.9 k	5.6 k
R10 & R11	0.47	0.47	0.47	0.47	0.47	0.33
R14	—	—	—	—	—	470 Ω
Q3	MPSA05	MPSA05	MPSU05 MJE180	MPSU05 MJE181	MPSU05 MJE181	MPSU05 MJE181
Q4	MPSA55	MPSA55	MPSU55 MJE170	MPSU55 MJE171	MPSU55 MJE171	MPSU55 MJE171
Q5	MJE170	MJE170	TIP42A	TIP42A	MJE2901T	MJE2901T
Q6	MJE180	MJE180	TIP41A	TIP41A	MJE2801T	MJE2801T
D1 & D2	MMD7000	MMD7000	MMD7000	MMD7000	MMD7000	MMD7000
*Heatsink	15.8°C/W	9.8°C/W	6.2°C/W	4.7°C/W	3.4°C/W	2°C/W

*Heatsink size calculation is based on a maximum ambient temperature of 50°C and a load phase angle of 60 degrees (see text for methods of calculation). The heat sink is for both devices on one sink.

FIGURE 4b – P.C. Board for 7 to 35 Watt Amplifier (NPN Predriver) Copper Side

TABLE 4a – Amplifier Performance Characteristics

Reference Figure 4	7 W 26 Vdc	10 W 30 Vdc	15 W 36 Vdc	20 W 43 Vdc	25 W 48 Vdc	35 W 54 Vdc
1. Idle Current (nominal no-signal)	1.6 mA	20 mA	28 mA	65 mA	68 mA	32 mA
2. Current Drain at Rated Power	425 mA	510 mA	610 mA	720 mA	320 mA	940 mA
3. Typical Input Impedance	230 kohms	320 kohms	230 kohms	230 kohms	230 kohms	220 kohms
4. THD at Rated Output Power 20 kHz or 1 kHz to 20 kHz	< 0.5%	< 0.5%	< 0.5%	< 0.5%	< 0.5%	< 0.5%
5. IM Distortion at 60 and 7000 Hz 4:1 ratio at Rated Power	<1%	<1%	<1%	<1%	<1%	<1%
6. -3 dB Bandwidth	16 Hz-300 kHz	15 Hz-380 kHz	16 Hz-250 kHz	16 Hz-250 kHz	16 Hz-275 kHz	16 Hz-230 kHz
7. Typical input sensitivity for rated output power	210 mV	260 mV	200 mV	220 mV	250 mV	280 mV
8. Maximum output power at 5% THD without current limiting	8.8 Watts	12.5 Watts	18 Watts	27.38 Watts	33.6 Watts	45.4 Watts
9. Maximum output power at 5% THD with current limiting	8.85 Watts	12.5 Watts	18 Watts	26.3 Watts	30 Watts	45 Watts
10. Power Supply Ripple Rejection	44.4 dB	42 dB	40 dB	30 dB	36 dB	34 dB
11. Short Circuit Power Supply Current with Current Limiting	1.32 Amps	1.45 Amps	1.5 Amps	1.5 Amps	1.62 Amps	2.2 Amps

7- to 35-watt Audio Amplifiers

Here are two circuits that show a general design for audio amplifiers that output from 7 to 35 watts. The first is an NPN design; the second PNP. Source: "Basic Design of Medium Power Audio Amplifiers," Motorola Semiconductor Products, Application Note AN-484A, Motorola, Inc.

1. AUDIO AMPLIFIERS

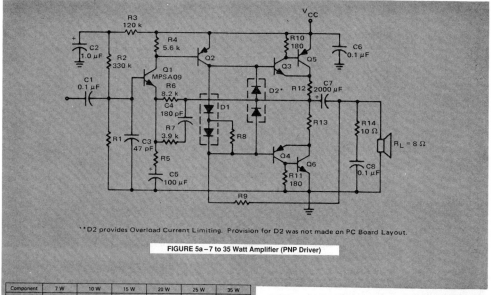

**D2 provides Overload Current Limiting. Provision for D2 was not made on PC Board Layout.

FIGURE 5a – 7 to 35 Watt Amplifier (PNP Driver)

Component	7 W	10 W	15 W	20 W	25 W	35 W
V_{CC}	25 V	30 V	38 V	46 V	48 V	54 V
R1	560 k	560 k	620 k	620 k	620 k	620 k
R5	100 Ω	82 Ω	100 Ω	100 Ω	150 Ω	180 Ω
R8	Value Selected to Provide 30 mA Collector Current in Q5.					
R9	4.7 k	4.7 k	8.2 k	8.2 k	8.2 k	8.2 k
R12 & R13	0.47	0.47	0.47	0.47	0.33	0.33
Q2	2N5087	2N5087	2N5087	2N5087	MPSA56	MPSA56
Q3	MPSA05	MPSA05	MPSU05	MPSU05	MPSU05	MPSU05
Q4	MPSA55	MPSA55	MPSU55	MPSU55	MPSU55	MPSU55
Q5			TIP42A	TIP42A	MJE2901T	MJE2901T
Q6			TIP41A	TIP41A	MJE2801T	MJE2801T
D1 & D2	MMD7000	MMD7000	MMD7000	MMD7000	MMD7000	MMD7000
*Heatsink	15.3°C/W	9.8°C/W	6.2°C/W	4.7°C/W	3.4°C/W	2°C/W

*Heatsink size calculation is based on a maximum ambient temperature of 50°C and a load phase angle of 60 degrees (see text for method of calculation). The heatsink is for both devices on one sink.

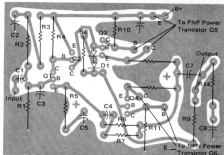

FIGURE 5b – P.C. Board, 7 to 35 Watt Amplifier

TABLE 5a – Amplifier Performance Characteristics

Reference Figure 5	7 W 25 Vdc	10 W 30 Vdc	15 W 28 Vdc	20 W 46 Vdc	25 W 48 Vdc	35 W 54 Vdc
1. Idle Current (nominal no-signal)	28 mA	40 mA	20 mA	58 mA	20 mA	37 mA
2. Current Drain at Rated Power	440 mA	500 mA	670 mA	720 mA	820 mA	940 mA
3. Typical Input Impedance	230 kohms	230 kohms	210 kohms	220 kohms	220 kohms	210 kohms
4. THD at Rated Output Power 20 Hz or 1 kHz to 20 kHz	<0.5%	<0.5%	<0.5%	<0.5%	<0.5%	<0.5%
5. IM Distortion at 60 and 7000 Hz 4:1 ratio at Rated Power	<1%	<1%	<1%	<1%	<1%	<1%
6. –3 dB Bandwidth	12 Hz-250 kHz	18 Hz-110 kHz	17 Hz-55 kHz	18 Hz-150 kHz	10 Hz-160 kHz	13 Hz-60 kHz
7. Typical input sensitivity for rated output power	220 mV	270 mV	110 mV	120 mV	230 mV	270 mV
8. Maximum output power at 5% THD without current limiting	8.8 Watts	12.5 Watts	21.7 Watts	27 Watts	34 Watts	43 Watts
9. Maximum output power at 5% THD with current limiting	8.75 Watts	11.5 Watts	21.1 Watts	24.5 Watts	34 Watts	43 Watts
10. Power Supply Ripple Rejection	37 dB	26 dB	41.94 dB	33 dB	36.5 dB	38 dB
11. Short Circuit Power Supply Current with Current Limiting	1.2 Amps	1.3 Amps	1.32 Amps	1.4 Amps	2 Amps	2 Amps

1. AUDIO AMPLIFIERS

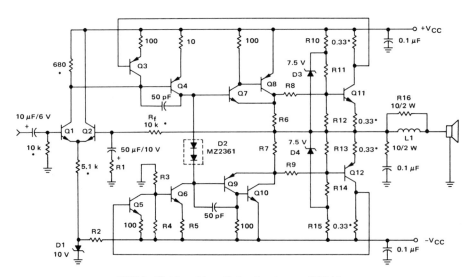

NOTE 1: All of the resistors with the values shown are ±10% tolerance, except where * indicates ±5%.

2: L1 is #20 wire close-wound for the full length of resistor, R16.

Output Power (Watts-rms)	Load Impedance (Ohms)	Output Transistors NPN (Q10)	Output Transistors PNP (Q8)	Driver Transistors NPN (Q7)	Driver Transistors PNP (Q9)	Pre-Driver Transistors NPN (Q6)	Pre-Driver Transistors PNP (Q4)	Differential Amplifier Transistors (Q1 & Q2)
35	4	2N5877	2N5875	MPSU05	MPSU55	MPSA05	MPSA55	MD8001
	8	MJE2801T	MJE2901T	MPSU05	MPSU55	MPSA06	MPSA56	MD8001
50	4	2N5302	2N4399	MPSU05	MPSU55	MPSA06	MPSA56	MD8001
	8	2N5878	2N5876	MPSU06	MPSU56	MPSA06	MPSA56	MD8002
60	4	2N5302	2N4399	MPSU06	MPSU56	MPSA06	MPSA56	MD8001
	8	2N5878	2N5876	MPSU06	MPSU56	MPSA06	MPSA56	MD8002
75	4	MJ802	MJ4502	MPSU06	MPSU56	MPSA06	MPSA56	MD8001
	8	MJ802	MJ4502	MM3007	2N5679	MM3007	MM4007	MD8003
100	4	MJ802	MJ4502	MPSU06	MPSU56	MPSU06	MPSU56	MD8002
	8	MJ802	MJ4502	MM3007	2N5679	MM3007	MM4007	MD8003

The following semiconductors are used at all of the power levels:

Q11 — MPSL01
Q5 — MPSA20
Q12 — MPSL51
Q3 — MPSA70

D1 — 1N5240A or 1N968A (See Note 1)
D2 — MZ2361
D3 & D4 — 1N5236B (See Note 1)

NOTE 1: For a low-cost zener diode, an emitter-base junction of a silicon transistor can be substituted. A transistor similar to the MPS6512 can be used for the 7.5 V zener.

1. AUDIO AMPLIFIERS

◀ High Power Audio Amplifier

For audio applications that require a little more "punch," this amplifier should fill the bill nicely. It can output from 35 to 100 watts depending on the components (as shown in the parts list). It can be built using low-cost, commonly available parts and has protection circuitry that allows it to operate safely under any usable load condition including a full short. Source: Richard G. Ruehs, "High Power Audio Amplifiers with Short Circuit Protection," Motorola Semiconductor Products, Application Note AN-485, Motorola, Inc.

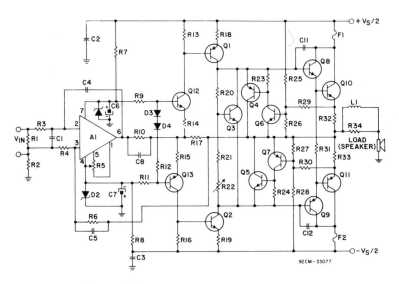

PARTS LIST FOR AMPLIFIER WITH 4 OR 8-OHM LOAD

Components	4 Ohms	8 Ohms
R1 (note 1)	10 K	10 K
R2	1	1
R3	1 K	1 K
R4	220	220
R5 (note 2)	Pot, 10 K	Pot, 10 K
R6	8.2 K	8.2 K
R7	1 K, 1W	1.8 K, 1W
R8	1 K, 1W	1.8 K, 1W
R9	1.8 K	1.8 K
R10	2.2 K	2.2 K
R11	1.8 K	1.8 K
R12	220	220
R13	4.7 K	1.8 K
R14	820	820
R15	820	820
R16	4.7 K	1.8 K
R17	39 K	39 K
R18	47	47
R19	47	47
R20	390	1 K
R21	56	56
R22 (note 3)	Pot, 1 K	Pot, 1 K
R23	100	100
R24	100	100
R25	3.9 K, 1W	8.2 K, 1W
R26	50	68
R27	50	68
R28	3.9 K, 1W	8.2 K, 1W
R29	180	470
R30	180	470
R31 (note 7)	100	100
R32	0.27, 7W	0.68, 7W
R33	0.27, 7W	0.68, 7W
R34	4.7, 1W	10, 1W

Components	4 Ohms	8 Ohms
C1	100pF	100pF
C2	0.47µF, 50V	0.47µF, 50V
C3	0.47µF, 50V	0.47µF, 50V
C4	12pF	12pF
C5	100pF	100pF
C6	22µF, 25V	22µF, 25V
C7	22µF, 25V	22µF, 25V
C8	10nF	10nF
C11 (note 7)	3.9nF	3.9nF
C12 (note 7)	3.9nF	3.9nF
D1	Zener, 15V	Zener 15V
D2	Zener, 15V	Zener 15V
D3	1N4148	1N4148
D4	1N4148	1N4148
Q1 (note 4)	RCA1A10	RCA1A10
Q2 (note 4)	RCA1A11	RCA1A11
Q3 (note 5)	RCA1A18	RCA1A18
Q4	RCP700A	RCP700A
Q5	RCP701A	RCP701A
Q6	RCA1A18	RCA1A18
Q7	RCA1A19	RCA1A19
Q8 (note 6)	RCA1C03	2N6474
Q9 (note 6)	RCA1C04	2N6476
Q10 (note 6)	BD751B	BD751C
Q11 (note 6)	BD750B	BD750C
Q12 (note 4)	RCA1A11	RCA1A11
Q13 (note 4)	RCA1A10	RCA1A10
A1	CA3100	CA3100
F1	4A	3A
F2	4A	3A
L1	2µH	4µH
V_S	78V	104V

Notes for Parts List

1. All resistors are non-inductive.
2. Adjust for an output of zero volts with zero volts at the input.
3. Adjust for a quiescent current of 200 mA.
4. Mount each device on heatsink of 30 cm^2 minimum area.
5. Mount on same heatsink as driver and output devices Q_8, Q_9, Q_{10} and Q_{11}.
6. Provide heatsinking as described in text.
7. These components cannot be found on the components layout of Fig. A-1. They are to be mounted directly on the driver-device sockets that are fixed on the heatsink.

1. AUDIO AMPLIFIERS

◀ 100-watt Audio Amplifier

This simple but effective audio amplifier consists of an IC input stage and a power stage that's composed of discrete transistors. It's essentially two cascaded-gain blocks with a single common feedback loop. The IC input stage offers a high unity-gain crossover frequency, wide power bandwidth, a high slew rate, low noise, and low offset. See the original application note for details on PC board and parts layout. Source: C. Raimarckers, "One-Hundred-Watt True-Complementary-Symmetry Audio Amplifier Using BD750 and BD751 Silicon Transistors," Power Transistors, Application Note AN-6904, RCA Solid State Division.

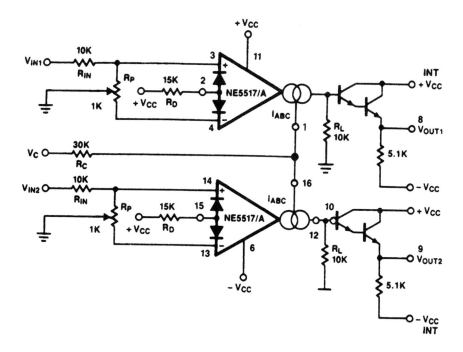

Gain-Controlled Stereo Amplifier

This stereo amplifier has variable gain that you can set from the control input. It uses a pair of Signetics NE5517A dual operational transconductance amplifiers. This amplifier has excellent tracking of up to 0.3 dB. You can adjust the offset with the R_p potentiometer. If this amplifier will be used for AC-coupled applications, you can replace the potentiometer with two 5.1-kohm resistors. Source: Linear Data Manual, Volume 2: Industrial, Signetics Corp.

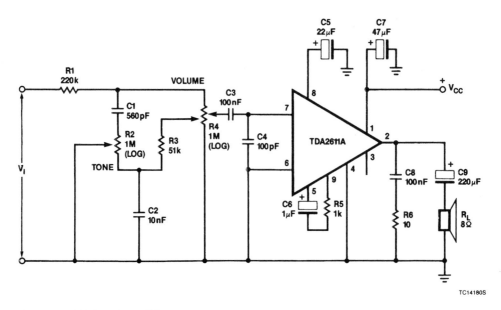

Ceramic Pickup Amplifier

Ceramic phono cartridges are still used in low-cost record-playing applications. This complete amplifier circuit takes the input from a ceramic cartridge and amplifies to an output of up to 5 watts. The heart of the circuit is the Signetics TDA2611A 5W audio amplifier. Source: "Linear Data Manual, Volume 1: Communications," Signetics Corporation.

Chapter 2

Amplifiers

Variable Gain Amplifier
Lock-In Amplifier
DC-Coupled Inverting Amplifier
DC-Coupled Noninverting Amplifier
Chopper-Stabilized Instrumentation Amplifier
± 5-volt Precision Instrumentation Amplifier
Precision High-Speed Op Amp
Large Signal-Swing Output Amplifier
Ultra-Precision Instrumentation Amplifier
Precision Isolation Amplifier
DC-Stabilized Low Noise Amplifier
AC-Coupled Inverting Amplifier
Current Mode Feedback Amplifier
Fast DC-Stablized FET Amplifier
Voltage-Controlled Amplifier
Fast-Stablized Noninverting Amplifier
Gain-Trimmable Wideband FET Amplifier
Low-Power Voltage Boosted Output Op Amp
Single-Supply Differential Bridge Amplifier
Three-Channel Separate-Gain Amplifier
N-Stage Parallel-Input Amplifier
Precision Amplifier with Notch
Differential Input/Output Amplifier

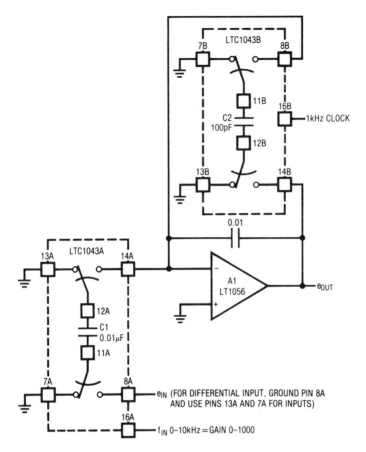

Variable Gain Amplifier

Designing a wide-range, digitally variable gain block with good DC stability is a difficult task. Normally, you need to use relays or temperature-compensated FET networks in expensive and complex arrangements. But this circuit uses two LTC1043 switched capacitor circuits to vastly simplify the design process. It has a continuously variable gain from 0–1000, gain stability of 20ppm/°C, and single-ended or differential input. Source: Jim Williams, "Applications for a Switched-Capacitor Instrumentation Building Block," Application Note 3, Linear Technology Corporation.

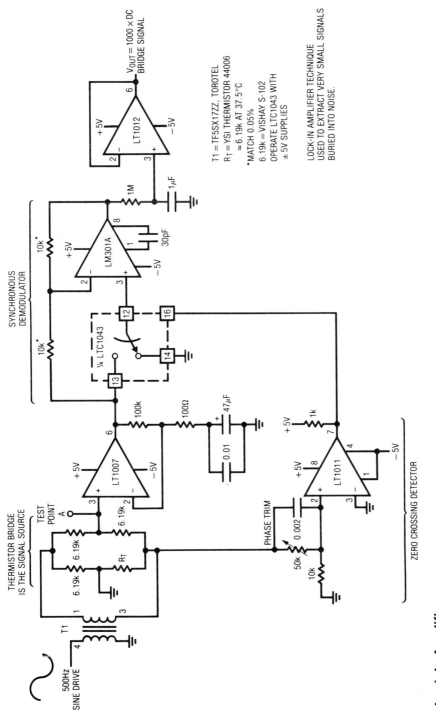

Lock-In Amplifier

A lock-in amplifier works by synchronously detecting the carrier-modulated output of the signal source. Because the desired signal information is contained within the carrier, the system constitutes an extremely narrow-band amplifier. The circuit rejects non-carrier-related components and passes only signals that are coherent with the carrier. Lock-in amplifiers can extract a signal up to 120 dB below the noise level. This design uses a thermistor bridge that detects extremely small temperature shifts in a biochemical microcalorimetry reaction chamber. Trim this circuit by adjusting the phase potentiometer so that C1 switches when the carrier crosses through zero. Source: Jim Williams, "Applications for a Switched-Capacitor Instrumentation Building Block," Application Note 3, Linear Technology Corporation.

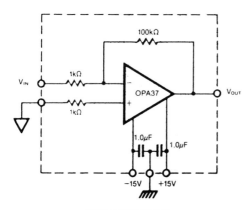
(A) Op Amp Realization

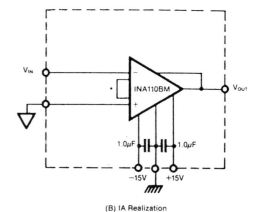
(B) IA Realization

*Pin-strapped for $A_v = 100$

	Op Amp Design	INA110 Design
Gain	100	100
Gain Error	2%	0.1%
Gain Tempco	100ppm/°C	20ppm/°C
V_{os}	60µV	280µV
V_{os} Drift	1.5µV/°C	2.5µV/°C
Gain Bandwidth	50MHz, typ	47MHz, typ
Settling Time (0.01%)	25µs, typ	4µs, typ
Input Impedance	1kΩ	∞

DC-Coupled Inverting Amplifier

You can drastically improve the performance of a garden-variety inverting amplifier by using an instrumentation amplifier instead of a standard op amp. The first circuit uses a Burr-Brown OPA37 op amp, and the second a Burr-Brown INA110BM instrumentation amplifier. This shows the improvement in gain-related errors, input impedance, and settling time. Source: "The Handbook of Linear IC Applications," Burr-Brown.

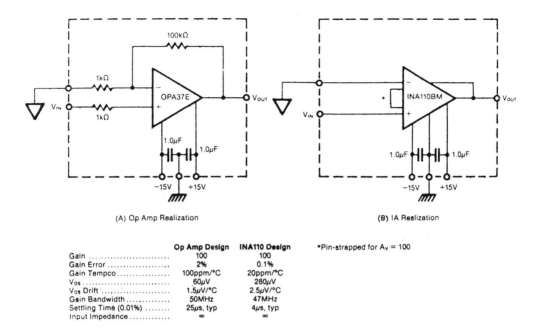

(A) Op Amp Realization (B) IA Realization

	Op Amp Design	INA110 Design
Gain	100	100
Gain Error	2%	0.1%
Gain Tempco	100ppm/°C	20ppm/°C
V_{os}	60µV	280µV
V_{os} Drift	1.5µV/°C	2.5µV/°C
Gain Bandwidth	50MHz	47MHz
Settling Time (0.01%)	25µs, typ	4µs, typ
Input Impedance	∞	∞

*Pin-strapped for $A_v = 100$

DC-Coupled Noninverting Amplifier

These two circuits illustrate the difference between using a standard op amp and an instrumentation amp for constructing a DC-coupled, non-inverting amplifier. The first circuit, constructed with an op amp, offers slightly better input and offsets characteristics, but poorer gain error and longer settling time than the instrumentation amplifier design. Source: "The Handbook of Linear IC Applications," Burr-Brown.

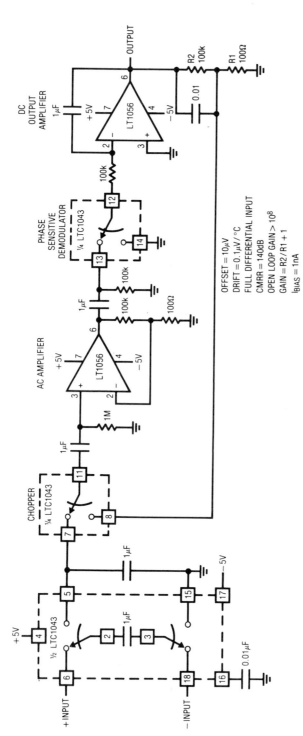

Chopper-Stabilized Instrumentation Amplifier

This low-drift chopper amplifier maintains true differential inputs while achieving 0.1μV / °C drift. Because the main amplifier is AC coupled, its DC terms do not affect overall circuit offset, resulting in extremely low offset and drift. Source: Jim Williams, "Applications for a Switched-Capacitor Instrumentation Building Block," Application Note 3, Linear Technology Corporation.

2. AMPLIFIERS

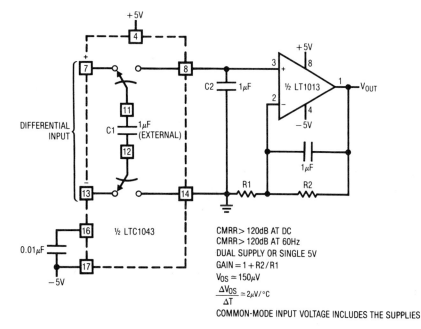

±5-volt Precision Instrumentation Amplifier

Using just two ICs, this circuit is a simple and economical way to build a high performance instrumentation amplifier. Its DC characteristics rival any IC or hybrid unit and it can operate from a single 5-V supply. The common-mode range includes the supply rails, allowing the circuit to read across shunts in the supply lines. The circuit's performance depends on the output amplifier. Although an LT11013 is shown, you can obtain lower figures for offset, drift, and bias current by using the Linear Technology LT1001, LT1012, LT1056, or LTC1052. Source: Jim Williams, "Applications for a Switched-Capacitor Instrumentation Building Block," Application Note 3, Linear Technology Corporation.

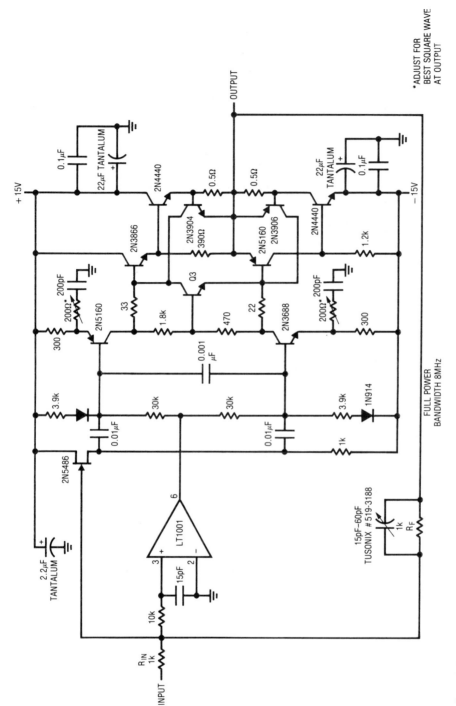

Precision High-Speed Op Amp

This 1000 V/μsec circuit meets the conflicting requirements of speed, accuracy, and output. It features a 1500 V/μsec slew rate, full output to 8 MHz, and will drive ± 10 V into a 10-ohm load. It's also short circuit protected at + 1 A. Because of this circuit's characteristics, RF layout techniques and a ground plane are mandatory, and the 2N4440s must be heat sunk. Source: Jim Williams, "Applications of New Precision Op Amps," Application Note 6, Linear Technology Corporation.

2. AMPLIFIERS

Large Signal-Swing Output Amplifier

If your power supply voltage is below 18 V DC, you can use either the RCA CA3020 or CA3020A for this circuit. For supply voltages up to 25 V DC with noninductive loads, you should use only the CA3020A. This circuit provides a gain of 60 dB and a bandwidth of 3.2 MHz if the output transistor Q_7 has a bypassed emitter resistor. With an unbypassed output emitter, the gain is 40 dB and the bandwidth is 8 MHz. The output stage can deliver a 5 V RMS signal with a supply voltage of 18 V DC. Source: W. M. Austin and H. M. Kleinman, "Application of the RCA CA3020 and CA3020A Integrated-Circuit Multi-Purpose Wide-Band Power Amplifiers," Application Note ICAN-5766, RCA Solid State.

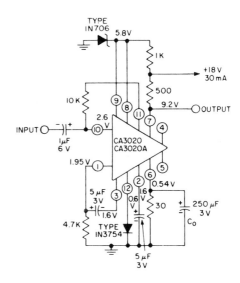

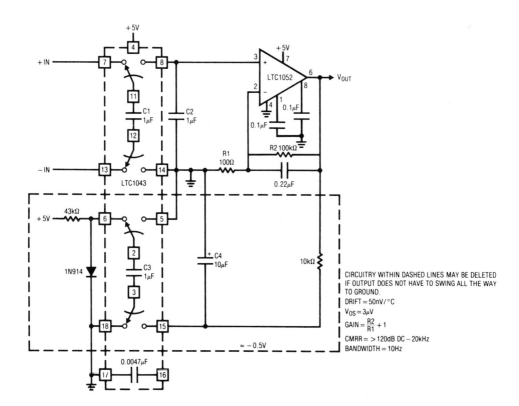

Ultra-Precision Instrumentation Amplifier

Besides offering greater accuracy and lower drift than any commercially available IC, hybrid, or module, this instrumentation amplifier will operate from a single 5-V power supply. Source: Jim Williams, "Application Considerations and Circuits for a New Chopper-Stabilized Op Amp," Application Note 9, Linear Technology Corporation.

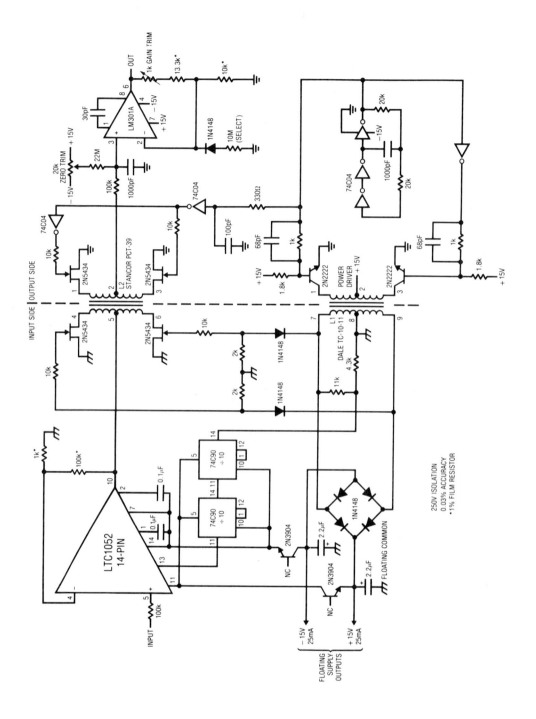

Precision Isolation Amplifier

Isolation amplifiers have inputs that are galvanically isolated from their output and power connections. This allows the amplifier to ignore the effects of ground loops and operate at input commonmode voltages many times the power supply voltage. This circuit provides a gain of 1000 and will operate at 250 V input common-mode levels. To trim this circuit, tie A1's input to floating common and adjust the zero trim for 0 V output. Next, connect A1's input to a +5 mV source and adjust the gain trim at A2 for exactly +5.000 V out. Finally, connect A1's input to a −5 mV source and select the 10 Mohm value in A2's feedback path for a −5.000 V output reading. Repeat this procedure until all three points are fixed. Source: Jim Williams, "Application Considerations and Circuits for a New Chopper-Stabilized Op Amp," Application Note 9, Linear Technology Corporation.

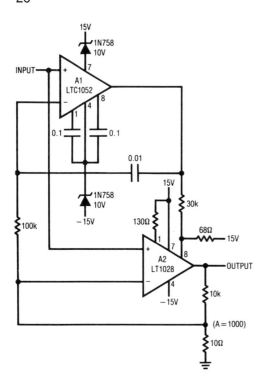

DC-Stabilized Low Noise Amplifier

This circuit shows a way to combine a low drift chopper stabilized with an ultra-low noise bipolar amplifier. Source: Jim Williams, "Composite Amplifiers," Application Note 21, Linear Technology Corporation.

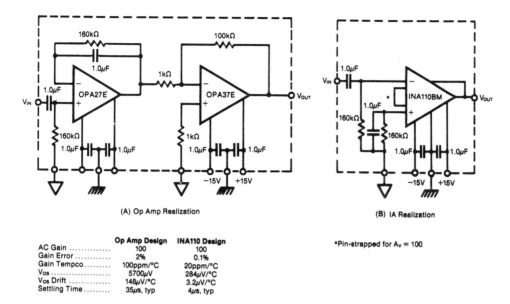

(A) Op Amp Realization (B) IA Realization

*Pin-strapped for $A_V = 100$

	Op Amp Design	INA110 Design
AC Gain	100	100
Gain Error	2%	0.1%
Gain Tempco	100ppm/°C	20ppm/°C
V_{os}	5700µV	284µV/°C
V_{os} Drift	148µV/°C	3.2µV/°C
Settling Time	35µs, typ	4µs, typ

AC-Coupled Inverting Amplifier

These two amplifiers, one built with op amps, the other with an instrumentation amplifier, both have a -3-dB cutoff at 1 Hz. Because of the high cost and space penalties of large-value capacitors, it's common to limit the capacitor to a smaller value, requiring a high-impedance device. Frequently the input impedance problem is addressed with the use of a unity-gain buffer, and the cost of this additional amplifier includes board space, power, and the cost of feedback and bypass components. The low value of cutoff frequency and small capacitor values also require larger values of input resistance. But as shown in the chart, this has a disastrous effect on the offset and offset drift of the amplifier. In this case, the instrumentation amplifier used in the second circuit provides a much cleaner solution. Source: "The Handbook of Linear IC Applications," Burr-Brown.

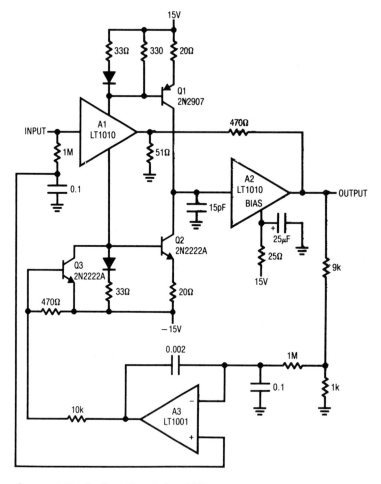

Current Mode Feedback Amplifier

This noninverting fast amplifier with wide output swing is based on an arrangement commonly referred to as "current mode feedback." This permits fixed bandwidth over a wide range of closed loop gains. This contrasts with normal feedback schemes, where bandwidth degrades as closed loop gain increases. For this circuit, full power bandwidth remains at 1 MHz over gains of between 1 and about 20. The LTY1010 buffers limit bandwidth in this circuit, and you can realize dramatic speed improvements if you replace them with discrete stages. Source: Jim Williams, "Composite Amplifiers," Application Note 21, Linear Technology Corporation.

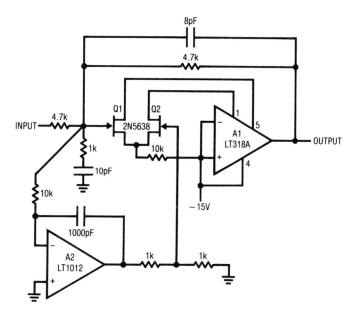

Fast DC-Stabilized FET Amplifier

In many practical applications, you need an amplifier that has extremely high performance in several areas, such as high speed and DC precision. If a single device can't do the job, you can configure a composite amplifier made up of two (or more) to do the job. Composite designs, such as this circuit, combine the best features of two or more amplifiers to achieve performance unatainable in a single device. But composite designs also permit circuit approaches that are normally impractical. This circuit uses discrete FETs for high speed. Source: Jim Williams, "Composite Amplifiers," Application Note 21, Linear Technology Corporation.

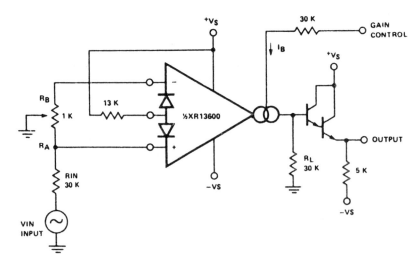

Voltage-Controlled Amplifier

This VCO is based around the EXAR XR-13600 dual operational transconductance amplifier, and shows how you can use linearizing diodes in a VCO application. The potentiometer is adjusted to minimize the effects of the control signal at the output. Source: EXAR Databook, EXAR Corporation.

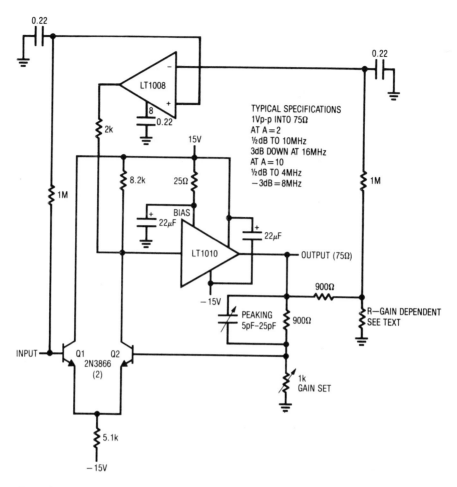

Fast-Stabilized Noninverting Amplifier

This DC-stabilized fast amplifier functions over a wide range of gains (typically 1 to 10), and is designed for fast applications where relatively low output swing is required. Its 1 Vp-p output works nicely for video circuits. The circuit combines the LT1010 and a fast discrete stage with an LT1008-based DC stabilizing loop. You should optimize the peaking adjustment under loaded output conditions. Source: Jim Williams, "Composite Amplifiers," Application Note 21, Linear Technology Corporation.

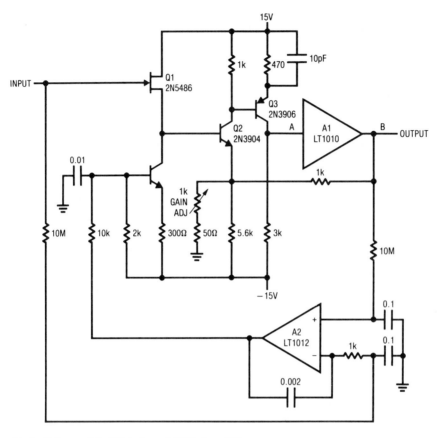

Gain-Trimmable Wideband FET Amplifier
This amplifier maintains high speed and low bias while achieving a true unity gain transfer function. With the optional discrete stage, slew exceeds 1000 V/μsec and full power bandwidth (1 Vp–p) is 18 MHz. −3 dB bandwidth is 58 MHz. Source: Jim Williams, "Composite Amplifiers," Application Note 21, Linear Technology Corporation.

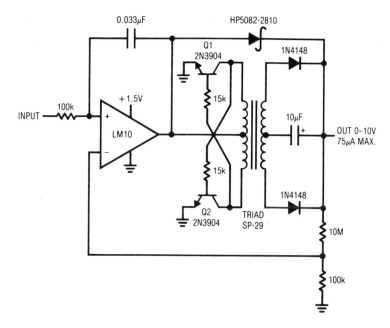

Low-Power Voltage-Boosted Output Op Amp

In many circumstances, it's desirable to have a 1.5-V powered circuit interface to a higher voltage system. An obvious example is a 1.5-V driven remote data-acquisition apparatus that feeds an AC line powered data gathering point. Although the battery powered portion can process signals locally, it's useful to be able to address the monitoring high-level instrumentation at high voltages. This 1.5 V–powered amplifier provides 0–10 V outputs at up to 75 μA capacity. In this case, the amplifier is set up with a gain of 101, though other gains are easy to design. Source: Jim Williams, "Circuitry for Single Cell Operation," Application Note 15, Linear Technology Corporation.

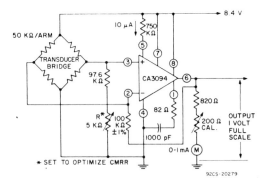

Single-Supply Differential Bridge Amplifier

In instrumentation applications, one of the more difficult (and frequently encountered) problems is the conversion of a differential input signal to a single-ended output signal. Although this conversion can be done in a straightforward way using op amps, the stringent matching requirements of resistor ratios in feedback networks make the conversion particularly difficult from a practical standpoint. RCA's CA-3094 is a monolithic programmable power switch/amplifier IC that consists of a high-gain preamplifier driving a power-output amplifier stage. Used in this circuit, it solves all the instrumentation problems. Source: L. R. Campbell and H. A. Wittlinger, "Some Applications of a Programmable Power Switch/Amplifier," Application Note ICAN-6048, RCA Solid State.

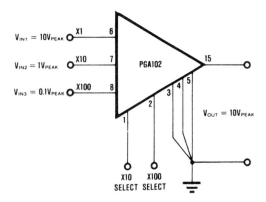

Three-Channel Separate-Gain Amplifier

As its name implies, this amplifier allows you to input three channels with ×1, ×10, ×100 gains. It uses the programmable-gain feature of the Burr-Brown PGA102 digitally controlled, fast-settling operational amplifier. Source: "Integrated Circuits Data Book," Burr-Brown.

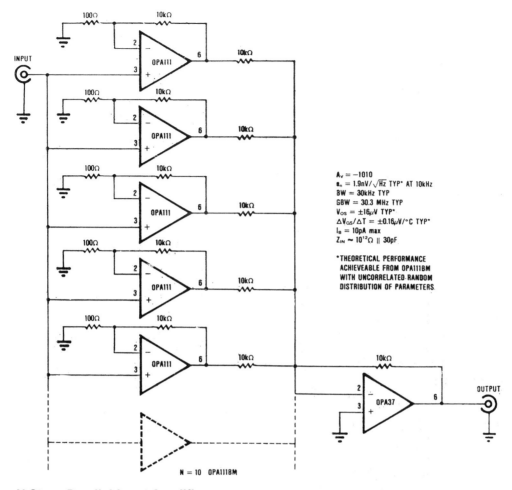

N-Stage Parallel-Input Amplifier

This N-Stage amplifier uses a series of Burr-Brown OPA111 precision monolithic dielectrically isolated FET (DIFET) operational amplifiers. The amplifier's specifications are as shown. Source: "Integrated Circuits Data Book," Burr-Brown.

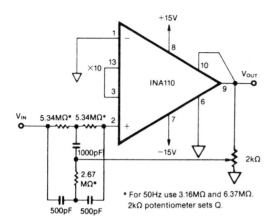

Precision Amplifier with Notch

This gain-of-10 amplifier includes a 60-Hz notch filter. It uses the Burr-Brown INA110 very high-accuracy instrumentation amplifier. Source: "Integrated Circuits Data Book," Burr-Brown.

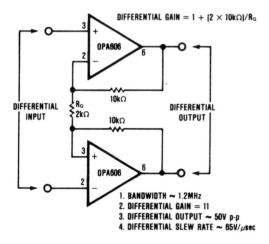

Differential Input/Output Amplifier

Burr-Brown's OPA606 is a wide-bandwidth monolithic dielectrically isolated FET op amp that has a wide bandwidth and low bias current. In this circuit, a pair of the ICs are used to construct a differential input/differential output amplifier with specifications as shown. Source: "Integrated Circuits Data Book," Burr-Brown.

Chapter 3
Audio Circuits

Microphone Preamplifier
Music Synthesizer
Voltage-Controlled Attenuator
Automatic Level Control
Tape Head Preamplifier
Variable-Slope Compressor–Expander
Hi-Fi Compandor
RIAA/NAB Compensation Preamplifier
Four-Input Stereo Source Selector
RIAA-Equalized Stereo Preamplifier
Fast Attack, Slow Release Hard Limiter
Audio Decibel Level Detector
Single-Frequency Audio Generator
AGC Amplifier
Audio Equalizer
Rumble/Scratch Filter
RIAA Preamplifier
Stereo Volume Control

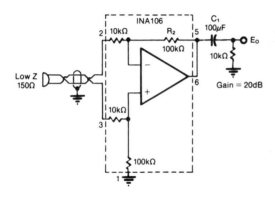

Microphone Preamplifier

Burr-Brown's INA106 is a precision, fixed-gain differential amplifier that consists of a premium-grade operation amplifier and a precision resistor network. In this circuit, it's used to make a simple and effective differential-input, low-impedance microphone preamplifier that will give a 20-dB gain. Source: "Integrated Circuits Data Book Supplement," Burr-Brown.

3. AUDIO CIRCUITS

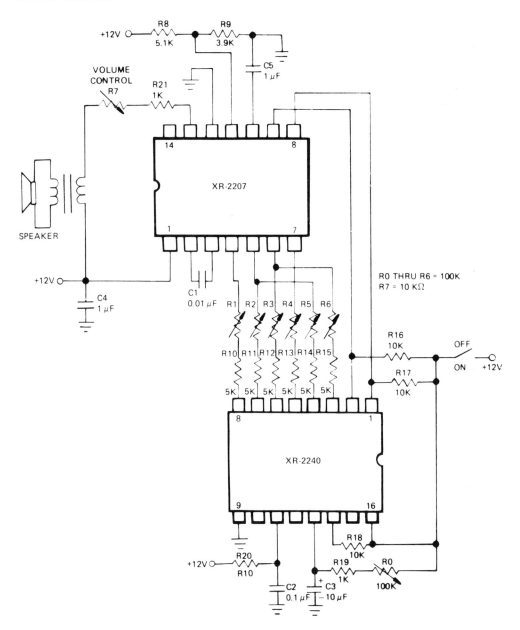

Music Synthesizer

This low-cost music synthesizer contains two monolithic ICs and a minimum number of external components. It's made up of the XR-2207 programmable tone generator IC, which is driven by the pseudo-random binary pulse pattern generated by the XR-2240 monolithic counter/timer circuit. Source: "An Electronic Music Synthesizer using the XR-2207 and the XR-2240," AN-15, EXAR Databook, EXAR Corporation.

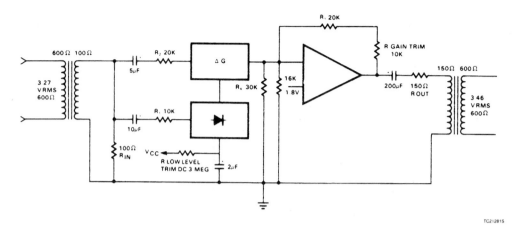

Voltage-Controlled Attenuator

The variable gain cell in Signetics' NE570/571 Compandor can be used as the heart of a high-quality voltage-controlled attenuator. With 0-V control voltage, R19 should be adjusted for 0-dB gain. At 1-V R9 should be adjusted for minimum distortion with a large (+10 dBm) input signal. The output DC bias should be measured at full attenuation and then R8 adjusted to give the same 0-dB gain. Properly adjusted, this circuit will give typically less than 0.1% distortion at any gain, with DC output voltage variation of only a few millivolts. This circuit can have a signal-to-noise ratio of up to 90 dB. If several of these units are required to track each other, a common exponential converter can be used. You can add transistors in parallel with Q2 to control the other channels. But for best tracking, the transistors should all be maintained at the same temperature. Source: "Applications for Compandors: NE570/571/SA571," AN174, — Linear Data Manual, Volume 1: Communications, Signetics Corporation.

3. AUDIO CIRCUITS

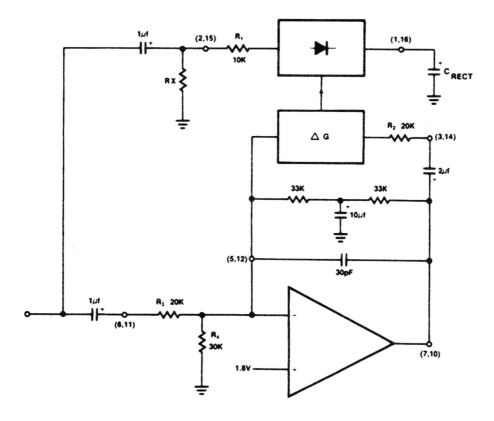

Automatic Level Control

This high-performance ALC uses Signetics' NE570 compandor. In this circuit, gain is inversely proportional to input level so that a 20-dB drop in input level will produce a 20-dB increase in gain. The output will remain fixed at a constant level. As shown, this circuit will maintain an output level of ±1 dB for an input range of +14 to −43 dB at 1KHz. Source: "Applications for Compandors: NE570/571/SA571," AN174, − Linear Data Manual, Volume 1: Communications, Signetics Corporation.

Tape Head Preamplifier

Today's low-noise recording tape formulations require low-noise electronics. For this tape playback head preamplifier, Burr-Brown's OPA37A ultra-low noise op amp is just the ticket. This preamp conforms to NAB (National Association of Broadcasters) specifications. Source: "Integrated Circuits Data Book," Burr-Brown.

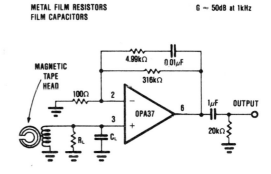

R_L AND C_L PER HEAD MANUFACTURER'S RECOMMENDATIONS

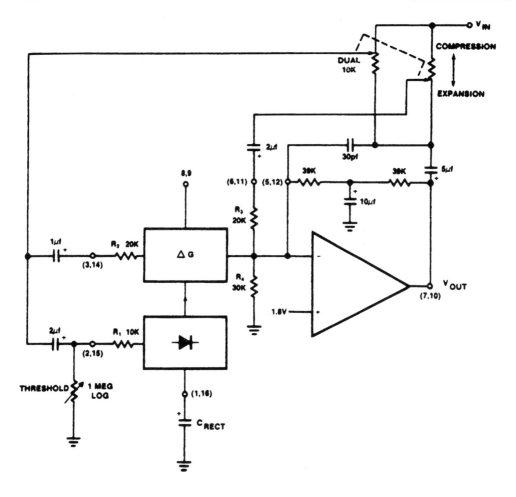

Variable Slope Compressor–Expander

This circuit, which uses the Signetics NE570 Compandor, lets you vary the compression and expansion ratios. Rotating the dual potentiometer causes the circuit hook-up to change from a basic compressor to a basic expander. In the center of rotation, the circuit ratio is 1:1, and it's continuously variable from 2:1 compression to 1:2 expansion. The optional threshold resistor will make the compression or expansion ratio deviate toward 1:1 at low levels. Source: "Applications for Compandors: NE570/571/SA571," AN174, — Linear Data Manual, Volume 1: Communications, Signetics Corporation.

3. AUDIO CIRCUITS

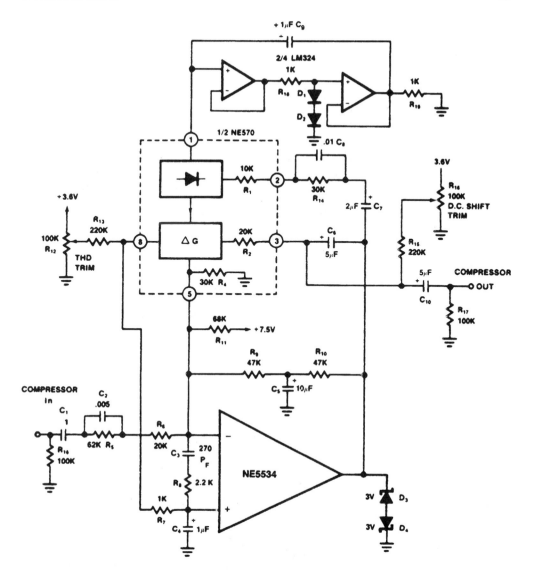

Hi-Fi Compandor

You can use NE570 compandor chips to construct a high-performance compandor that's suitable for music use. It can be used for noise reduction in tape recorders, transmission systems, bucket brigade delay lines, and digital audio delay lines. The two circuits (compressor on this page and expander on following page) shown here have features that improve performance. The compressor circuit has a high-frequency pre-emphasis and the expander has de-emphasis eliminating the problem of "breathing," where the change in background noise level would be otherwise heard as the system changes gain. See the original application note for a detailed discussion of customizing these circuits.
Source: "Applications for Compandors: NE570/571/SA571," AN174, — Linear Data Manual, Volume 1: Communications, Signetics Corporation.

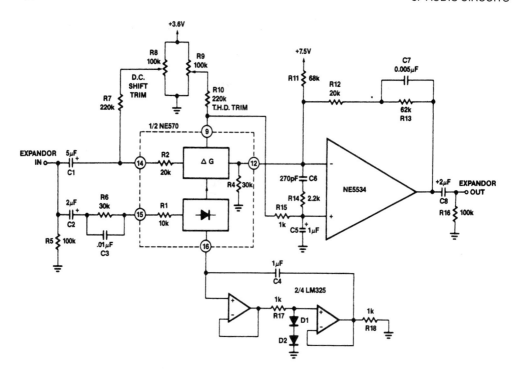

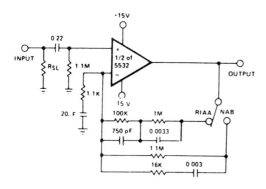

*SELECT TO PROVIDE SPECIFIED TRANSDUCER LOADING
OUTPUT NOISE 0.8 mV rms (WITH INPUT SHORTED)
ALL RESISTOR VALUES ARE IN OHMS

RIAA/NAB Compensation Preamplifier

Being able to switch between RIAA (Record Industry Association of America) and NAB (National Association of Broadcasters) equalizations can be useful in equipment that's used for both broadcast and nonbroadcast use. This simple preamplifier, built around an EXAR XR-5532 dual low-noise op amp, lets you switch between these two industry-standard equalizations. Since each channel uses one-half of the op amp, it's easy to adapt into a stereo circuit. Source: EXAR Databook, EXAR Corporation.

3. AUDIO CIRCUITS

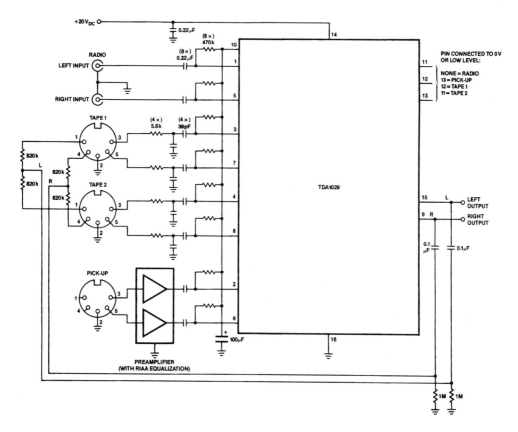

Four-Input Stereo Source Selector
This simple switch for selecting from four stereo sources is based on Signetics TDA1029 Stereo Audio Switch IC, a dual operational amplifier. Each amplifier has four mutually-switchable inputs that are protected by clamping diodes. The input currents are independent of switch position, and the outputs are short-circuit protected. Source: "Linear Data Manual, Volume 1: Communications," Signetics Corporation.

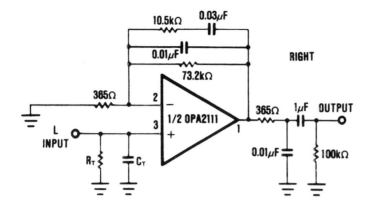

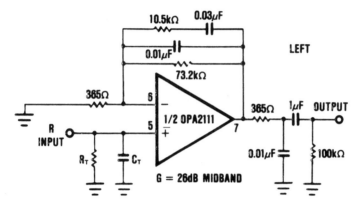

RIAA-Equalized Stereo Preamplifier

Using just one Burr Brown OPA211 DIFET op amp, this circuit can be used as a preamplifier for a wide range of audio sources. It conforms to RIAA specifications. Source: "Integrated Circuits Data Book," Burr-Brown.

3. AUDIO CIRCUITS

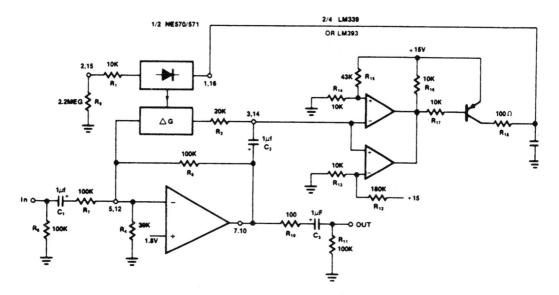

15V Pin 13
ND Pin 4
R1, R2, R4 are internal to the NE570/571.

Fast Attack, Slow Release Hard Limiter

Used for limiting audio levels, this limiter runs at nearly unity gain for small signals. But when the output signal tries to exceed a ±1-V peak, a comparator threshold is exceeded. The PNP transistor is turned on and rapidly charges C4, which activates the G cell. Negative feedback through the G cell reduces the gain and the output signal level. The attack time is set by the RC product of R18 and C4, and the release time is determined by C4 and the internal rectifier resistor, which is 10K. As shown here, the limiter attacks in less than 1 msec and has a release time constant of 100 msec. You can build a stereo limiter out of one NE570/571, one LM339, and two PNP transistors. The resistor network R12, R13, R14, and R15, which sets the limiting threshold, can be common between channels. To gang the stereo channels together, pins 1 and 16 should be jumpered together. In this situation the limiting in one channel will produce a corresponding gain in the second channel to maintain the balance of the stereo image. You can tie off all 4 comparators together, and you only need to use one PNP transistor and one capacitor (C4). Source: "Applications for Compandors: NE570/571/SA571," AN174, — Linear Data Manual, Volume 1: Communications, Signetics Corporation.

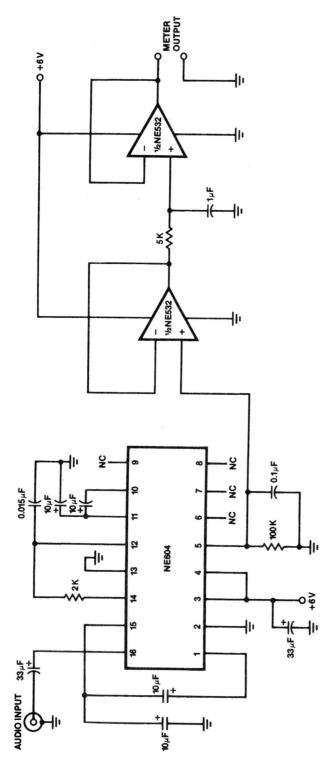

Audio Decibel Level Detector

Signetics NE604 was designed for cellular radio applications. But it has features that make it useful for other applications. The Received Signal Strength Indicator (RSSI) is necessary in cellular radio for monitoring of received signal strength by the radio's microcomputer. Using the NE604 along with an NE532 amplifier you can construct this audio level indicator that will drive a meter. Because it draws very little power (less than 5mA with a single 6V supply), it's ideal for portable battery-powered equipment. It has an 80dB dynamic range and 10.5uV sensitivity. Source: Robert J. Zavrel, Jr., "Audio Decibel Level Detector with Meter Driver," AN1991, Linear Data Manual, Volume 1: Communications, Signetics Corporation.

3. AUDIO CIRCUITS

Single-Frequency Audio Generator

A standard 555 timer is ideal for constructing a single-frequency audio generator. Use the equation shown to choose the R/C combination needed for the desired frequency. Source: "NE555 and NE556 Applications," AN170, Linear Data Manual Volume 2: Industrial, Signetics Corporation.

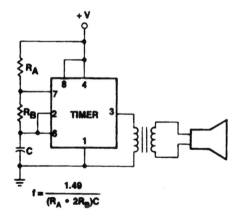

$$f = \frac{1.49}{(R_A + 2R_B)C}$$

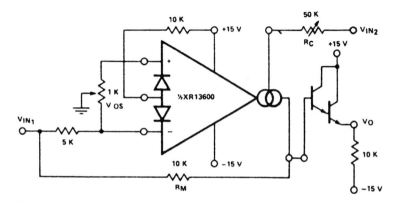

AGC Amplifier

This simple automatic gain control uses the EXAR XR-13600 dual operational transconductance amplifier. It's useful in a wide range of applications. Source: EXAR Databook, EXAR Corporation.

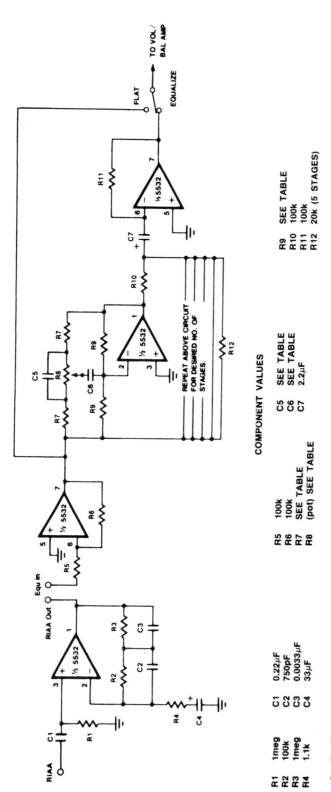

Audio Equalizer

The Signetics 5532 high-performance op amp is a good choice for use in high quality and professional audio equipment that requires low noise and low distortion. This audio equalizer consists of an RIAA preamplifier, input buffer, 5-band equalizer, and mixer. The RIAA preamplifier section is a standard compensation configuration with low-frequency boost provided by the magnetic cartridge and the RC network in the op amp feedback loop. Cartridge loading is accomplished via R1. 47K was chosen as a typical value, and may differ from cartridge to cartridge. The equalizer section consists of an input buffer, five active variable band pass/notch (depending on R9's setting) filters, and an output summing amplifier. Linear pots are recommended for R9. See the component value table for center frequencies and associated capacitor values. Source: "Audio Circuits Using the NE5532/33/34," AN142, Linear Data Manual Volume 2: Industrial, Signetics Corporation.

COMPONENT VALUES FOR FIGURE 1

R8 = 25k R7 = 2.4k R9 = 240k			R8 = 50k R7 = 5.1k R9 = 510k			R8 = 100k R7 = 10k R9 = 1meg		
f_o	C5	C6	f_o	C5	C6	f_o	C5	C6
23Hz	1µF	0.1µF	25Hz	0.47µF	0.047µF	12Hz	0.47µF	0.047µF
50Hz	0.47µF	0.047µF	36Hz	0.33µF	0.033µF	18Hz	0.33µF	0.033µF
72Hz	0.33µF	0.033µF	54Hz	0.22µF	0.022µF	27Hz	0.22µF	0.022µF
108Hz	0.22µF	0.022µF	79Hz	0.15µF	0.015µF	39Hz	0.15µF	0.015µF
158Hz	0.15µF	0.015µF	119Hz	0.1µF	0.01µF	59Hz	0.1µF	0.01µF
238Hz	0.1µF	0.01µF	145Hz	0.082µF	0.0082µF	72Hz	0.082µF	0.0082µF
290Hz	0.082µF	0.0082µF	175Hz	0.068µF	0.0068µF	87Hz	0.068µF	0.0068µF
350Hz	0.068µF	0.0068µF	212Hz	0.056µF	0.0056 µF	106Hz	0.056µF	0.0056µF
425Hz	0.055µF	0.0056µF	253Hz	0.047µF	0.0047µF	126Hz	0.047µF	0.0047µF
506Hz	0.047µF	0.0047µF	360Hz	0.033µF	0.0033µF	180Hz	0.033µF	0.0033µF
721Hz	0.033µF	0.0033µF	541Hz	0.022µF	0.0022µF	270Hz	0.022µF	0.0022µF
1082Hz	0.022µF	0.0022µF	794Hz	0.015µF	0.0015µF	397Hz	0.015µF	0.0015µF
1588Hz	0.015µF	0.0015µF	1191Hz	0.01µF	0.001µF	595Hz	0.01µF	0.001µF
2382Hz	0.01µF	0.001µF	1452Hz	0.0082µF	820pF	726Hz	0.0082µF	820pF
2904Hz	0.0082µF	820pF	1751Hz	0.0068µF	680pF	875Hz	0.0068µF	680pF
3502Hz	0.0068µF	680pF	2126Hz	0.0056µF	560pF	1063Hz	0.0056µF	560pF
4253Hz	0.0056µF	560pF	2534Hz	0.0047µF	470pF	1267Hz	0.0047µF	470pF
5068Hz	0.0047µF	470pF	3609Hz	0.0033µF	330pF	1804Hz	0.0033µF	330pF
7218Hz	0.0033µF	330pF	5413Hz	0.0022µF	220pF	2706Hz	0.0022µF	220pF
10827Hz	0.0022µF	220pF	7940Hz	0.0015µF	150pF	3970Hz	0.0015µF	150pF
15880Hz	0.0015µF	150pF	11910Hz	0.001µF	100pF	5955Hz	0.001µF	100pF
23820Hz	0.001µF	100pF	14524Hz	820pF	82pF	7262Hz	820pF	82pF
			17514Hz	680pF	68pF	8757Hz	680pF	68pF
			21267Hz	560pF	56pF	10633Hz	560pF	56pF
						12670Hz	470pF	47pF
						18045Hz	330pF	33pF

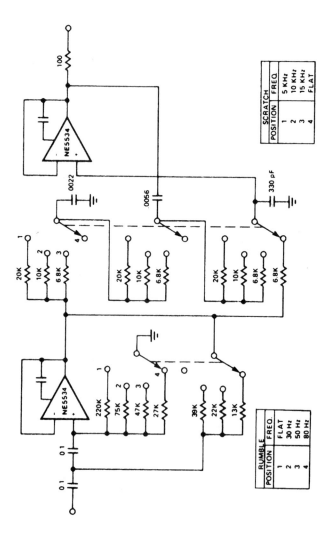

NOTE:
All resistor values are in ohms.

Rumble/Scratch Filter

A filter that removes low-frequency rumble and the high-frequency components of scratches can drastically improve audio quality. This filter, designed with op amps, uses the 2-pole Butterworth approach and features switchable break points. With this circuit, you can choose any degree of filtering from fairly sharp to virtually none at all. Source: "Audio Circuits Using the NE5532/33/34," AN142, Linear Data Manual Volume 2: Industrial, Signetics Corporation.

3. AUDIO CIRCUITS

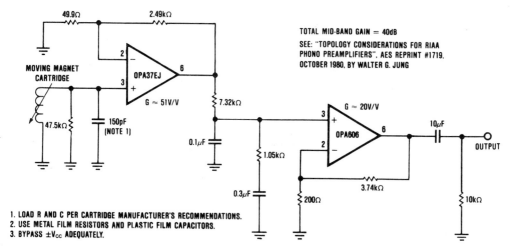

1. LOAD R AND C PER CARTRIDGE MANUFACTURER'S RECOMMENDATIONS.
2. USE METAL FILM RESISTORS AND PLASTIC FILM CAPACITORS.
3. BYPASS ±V$_{cc}$ ADEQUATELY.

RIAA Preamplifier

This low-noise, low-distortion preamplifier is simple and inexpensive to build, and can be used with any standard moving-magnet phono cartridge. It also conforms to RIAA specs. Source: "Integrated Circuits Data Book," Burr-Brown.

$$\frac{V_O}{V_{IN}} = 940 \times I_B \text{ (mA)}$$

Stereo Volume Control

This circuit uses the matching of the two EXAR XR-13600 dual operational transconductance amplifiers to provide a stereo volume control with a typical channel-to-channel gain tracking of 0.3 dB. As shown, R_P minimizes the output offset voltage. In AC-coupled applications, you can replace it with two 510-ohm resistors. Source: EXAR Databook, EXAR Corporation.

Chapter 4
Automotive Circuits

Tachometer
Auto Burglar Alarm
Speed Warning Device
Automobile Voltage Regulator
AM/FM Car Radio
AM Car Radio
6-watt Audio Amplifier

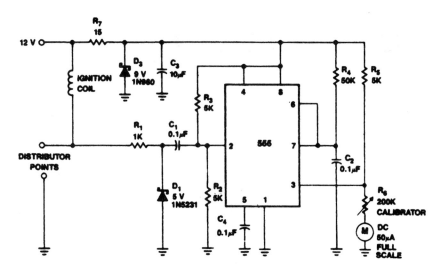

NOTE:
All resistor values are in ohms.

Tachometer

This automotive tachometer is built around a standard 556 dual timer. It receives its pulses from an auto's distributor points. Meter M receives a calibrated current through R6 when the timer output is high. After time-out, the meter receives no current for that part of the duty cycle. Integration of the variable duty cycle by the meter movement provides a visible indication of engine speed. Source: Linear Data Manual, Volume 2: Industrial, Signetics Corporation.

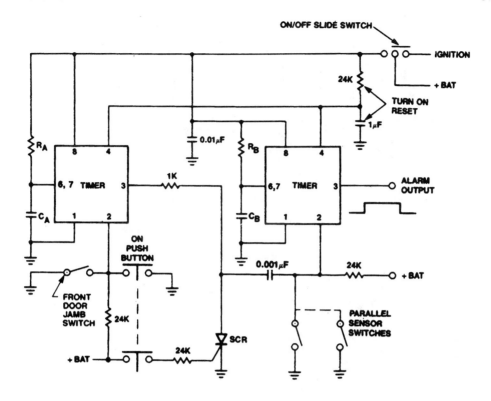

NOTES:
Timer Signetics NE555
All resistor values in ohms

Auto Burglar Alarm

A pair of industry-standard 555 timers provide the basis for a simple burglar alarm for automotive applications. Timer A produces a safeguard delay, allowing the driver to disarm the alarm and eliminating a vulnerable outside control switch. The SCR prevents timer A from triggering timer B unless timer B is triggered by strategically located sensor switches. Source: "NE555 And NE556 Applications," AN170, Linear Data Manual Volume 2: Industrial, Signetics Corporation.

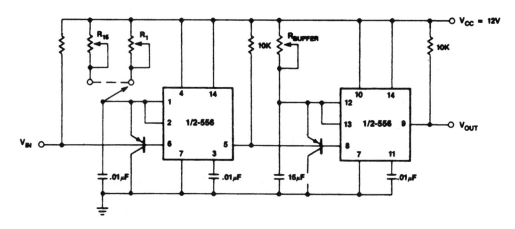

a. Schematic of Speed Warning Device

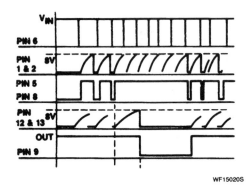

b. Operating Waveforms Speed Warning Device

Speed Warning Device

This simple circuit using a 556 dual timer makes an effective speed warning circuit. It uses the "missing pulse detector" concept. Source: Linear Data Manual Volume 2: Industrial, Signetics Corporation.

4. AUTOMOTIVE CIRCUITS

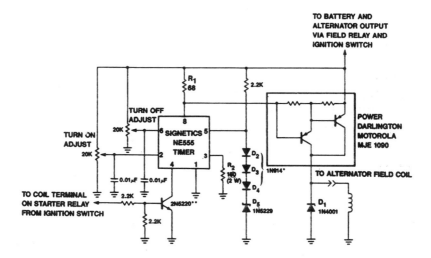

NOTES:
* Can be any general purpose Silicon diode or 1N4157 device.
** Can be any general purpose Silicon transistor.
All resistor values are in ohms.

Automobile Voltage Regulator
A monolithic 555-type timer is the heart of this simple automobile voltage regulator. When the timer is off so that its output (pin 3) is low, the power Darlington transistor pair is off. If battery voltage becomes too low (14.4V in this case), the timer turns on and the Darlington pair conducts. Source: "NE555 And NE556 Applications," AN170, Linear Data Manual, Volume 2: Industrial, Signetics Corporation.

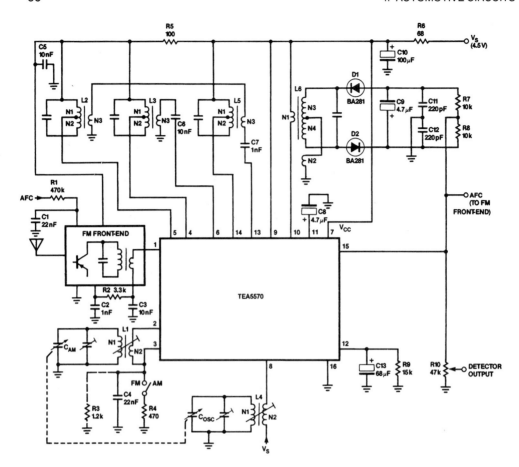

```
Coil Data
L2   N1  =  3
     N2  =  8
     N3  =  1
     C   =  82pF
L3   N1  =  33
     N2  = 113
     N3  =  9
     C   = 180pF
L4   N1  =  90
     N2  =  6
L5   N1  =  33
     N2  = 113
     N3  =  9
L6   N1  =  50
     N2  =  50
     N3  =  4.5
     N4  =  6.5
     C   =  82pF
```

AM/FM Car Radio

This AM/FM radio, primarily intended for automotive applications, is built around the Signetics TEA AM/FM radio receiver circuit. In addition to AM/FM switching, the IC incorporates for AM a double-balanced mixer, a one-pin oscillator, IF amplifier with AGC and detector, and a level detector for tuning indication. The FM circuitry comprises IF stages with a symmetrical limiter for a ratio detector. A level detector for mono/stereo switch information and/or indications completes the FM part of the chip. Source: "Linear Data Manual, Volume 1: Communications," Signetics Corporation.

4. AUTOMOTIVE CIRCUITS

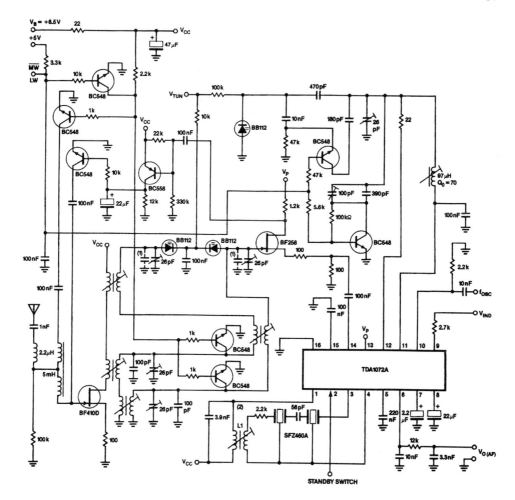

NOTES:
1. Values of capacitors depend on the selected group of capacitive diodes BB112.
2. For IF filter and coil data refer to Block Diagram.
3. The circuit includes pre-stage AGC optimized for good large-signal handling.

AM Car Radio

This single-chip AM car radio is built around Signetics' TDA1072A AM receiver circuit. The IC performs the active function and part of the filtering function of an AM radio receiver. Source: "Linear Data Manual, Volume 1: Communications," Signetics Corporation.

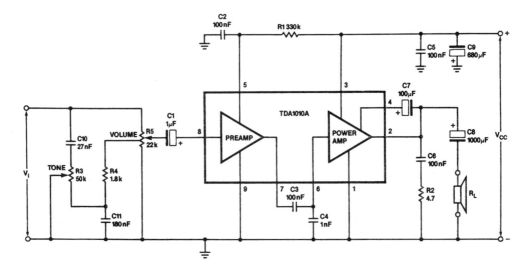

6-watt Audio Amplifier

This circuit draws its simplicity from the fact that the Signetics TDA1010A IC is an integrated 6-W amplifier that's complete with a preamplifier. Although useful in a wide range of low-power amplification applications, it was primarily developed as a car radio amplifier for use with 2- and 4-ohm applications. Its operating temperature range of −25°C to +150°C also makes it ideal for automotive applications. Source: "Linear Data Manual, Volume 1: Communications," Signetics Corporation.

Chapter 5

Battery Circuits

6-volt Battery Charger
12-volt Battery Charger
Wind-Powered Battery Charger
Battery Backup Regulator
Battery Splitter
High-Current Battery Splitter
Battery Cell Monitor
Micropower Switching Regulator
Sine Wave Output Converter
Voltage Doubler
Low-Power Flyback Regulator
Battery Monitor for High-Voltage Charging Circuits
Thermally Controlled NiCad Battery Charger
Single-Cell Up Converter
Low Dropout 5-volt Regulator
Computer Battery Control

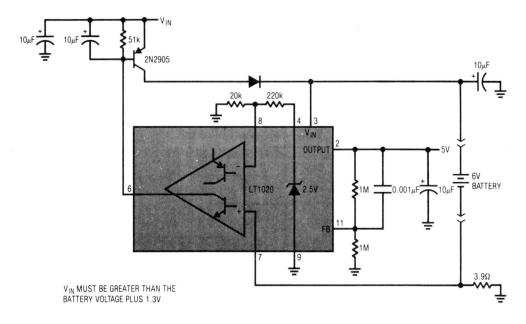

V_{IN} MUST BE GREATER THAN THE BATTERY VOLTAGE PLUS 1.3V

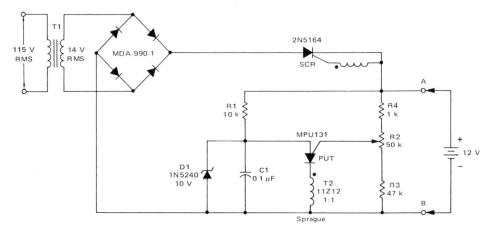

12-volt Battery Charger

This short-circuit-proof battery charger will provide an average charging current of about 8 amps to a 12-V lead acid storage battery. It has an additional advantage in that it won't operate (and won't be damaged) if the battery is improperly connected to the circuit. The charging voltage can be set from 10 to 14 V, with the lower limit set by D1 and the upper limit by T1. You can also obtain lower charging voltages by reducing the reference voltage (reducing the value of zener diode D1) and limiting the charging current (using either a lower voltage transformer (T1) or adding resistance in series with the SCR). Source: R. J. Haver and B. C. Shiner, "Theory, Characteristics and Applications of the Programmable Unijunction Transistor," Motorola Semiconductor Products, Application Note AN-527, Copyright Motorola, Inc. Used by permission.

5. BATTERY CIRCUITS

◄ 6-volt Battery Charger

This circuit will charge a standard rechargeable 6-V battery. It uses the LT1020 micropower regulator and comparator. With only 40-μA supply current, the IC can supply over 125 mA of output current. For this application, the input voltage should 1.3 V higher than the battery voltage (i.e., 7.3 V). Source: "Linear Databook Supplement," Linear Technology Corporation.

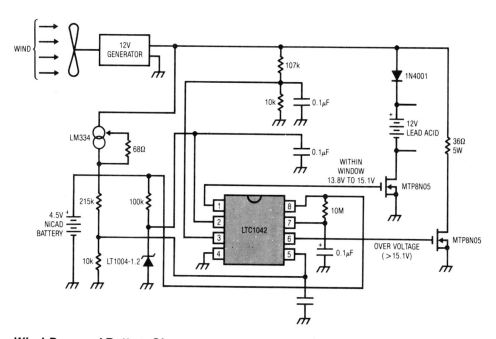

Wind-Powered Battery Charger

You can construct a simple, wind-powered battery charger using the Linear Technology LTC1042 window comparator, a 12-V DC permanent magnet motor, and a low-cost power FET transistor. This charger can be used as remote source of power where wind energy is plentiful such as on sailboats or at remote transmitter sites. Unlike solar-powered panels, this system will function in bad weather and at night. The DC motor is used as a generator with the voltage output proportional to its RPM. The LTC1042 monitors the voltage output and provides the following control functions: (1) If generator voltage output is below 13.8 V, the control circuit is active and the NiCad battery is charging through the LM334 current source. The lead acid battery is not being charged. (2) If the generator voltage output is between 13.8 and 15.1 V, the 12-V lead acid battery is being charged at about a 1 amp/hour rate, limited by the power FET. (3) If generator voltage exceeds 15.1 V (a condition caused by excessive wind speed of the 12-V battery being fully charged) then a fixed load is connected to limit the generator RPM and prevent damage. Source: "Linear Databook Supplement," Linear Technology Corporation.

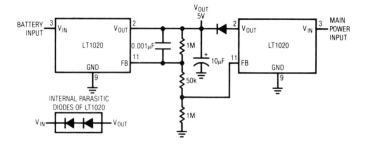

Battery Backup Regulator

This is a low-loss way to accomplish a "glitchless" memory battery backup. During line-powered operation, the left LT1020 does the work. The feedback string is arranged so that the right LT1020 doesn't conduct under line-powered conditions. When the AC line goes down, the associated LT1020 begins to go off, allowing the battery-driven regulator to turn on and maintain the load. Source: Jim Williams, "Micropower Circuits for Signal Conditioning," Application Note 23, Linear Technology Corporation.

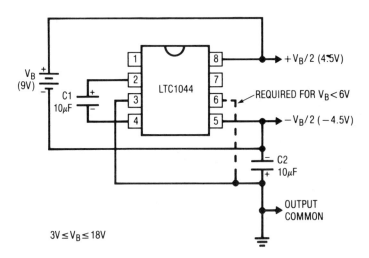

Battery Splitter

Obtaining both positive and negative supplies from a single battery is a common need in many applications. If your current requirements are small, this circuit is a simple way to do it. It provides symmetrical ± output voltages, each equal to one-half the input voltage. Source: Jim Williams, "Power Conditioning Techniques for Batteries," Application Note 8, Linear Technology Corporation.

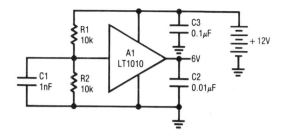

High-Current Battery Splitter

For applications that require high current, getting both positive and negative supplies from a battery requires a buffer, supplied by the LT1010 in this circuit. The circuit can source or sink up to ±150 mA with only 5-mA quiescent current. You can make the C2 output capacitor as large as necessary to absorb current transients. The input capacitor is used on the buffer to avoid high-frequency instability that can be caused by high source impedance. Source: Jim Williams, "Power Conditioning Techniques for Batteries," Application Note 8, Linear Technology Corporation.

Battery Cell Monitor

Rechargeable batteries are often charged in series, and it's important to be able to monitor the condition of each cell. This circuit uses the Burr-Brown INA117 precision unity-gain differential amplifier for that application. The circuit's operating range is up to ±200 V, and differential fault conditions in this range won't damage the amplifier. Also, since the INA117 doesn't require isolated front-end power, the cost per cell is low. Source: "Integrated Circuits Data Book Supplement," Burr-Brown.

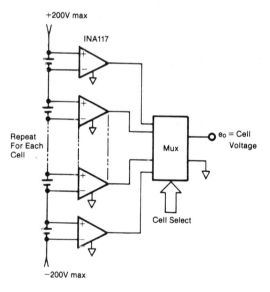

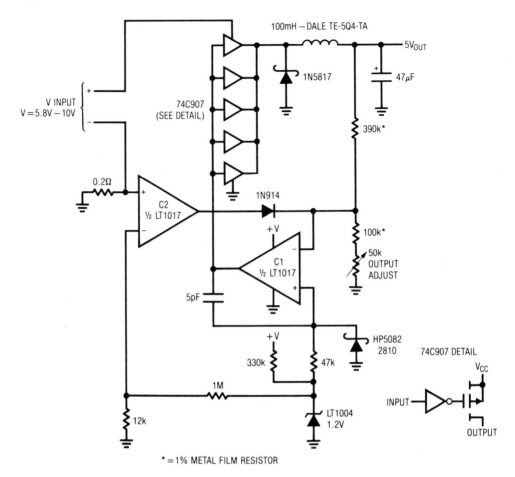

Micropower Switching Regulator

In many cases, it's necessary to efficiently convert battery-level voltages to different potentials to meet circuit requirements. This buck-type switching regulator has a quiescent drain of 70 µA, up to 90% efficiency, and an output current capability of up to 20 mA. This circuit is unusual in that it uses a CMOS buffer as a pass switch. Source: Jim Williams, "Micropower Circuits for Signal Conditioning," Application Note 23, Linear Technology Corporation.

5. BATTERY CIRCUITS

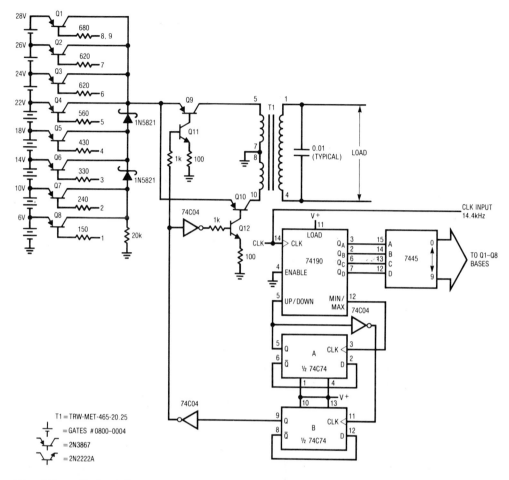

T1 = TRW-MET-465-20.25
= GATES #0800-0004
= 2N3867
= 2N2222A

Sine Wave Output Converter

Not all battery converters must produce a DC output. Some battery-driven systems use high voltage, sine-wave driven devices such as small motors, gyros, or synchros. Although you can derive high voltage sine waves from a battery using linear techniques, the efficiency is usually poor. This circuit obtains 78% efficiency by sequentially switching segments of a 28-V battery stack into a fixed gain, step-up voltage chopper. Source: Jim Williams, "Power Conditioning Techniques for Batteries," Application Note 8, Linear Technology Corporation.

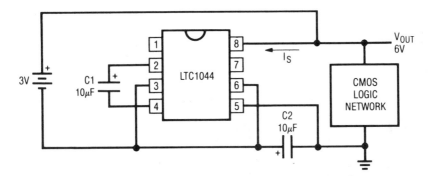

Voltage Doubler

Many applications call for a voltage that's higher than the battery output. This circuit is a simple way to double voltage using Linear Technology's LTC1044 switched-capacitor voltage converter. Source: Jim Williams, "Power Conditioning Techniques for Batteries," Application Note 8, Linear Technology Corporation.

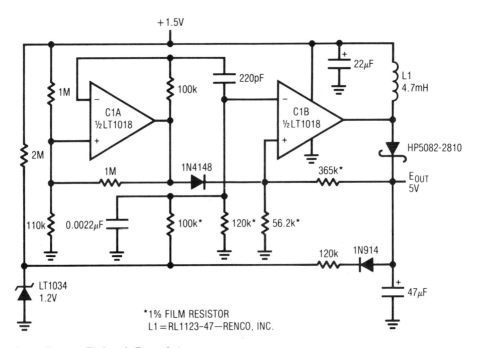

Low-Power Flyback Regulator

This circuit shows a way to let you use standard logic functions from a 1.5-V battery. It's a switching regulator, in a flyback configuration, specifically designed for low-power operation. Source: Jim Williams, "Circuitry for Single Cell Operation," Application Note 15, Linear Technology Corporation.

5. BATTERY CIRCUITS

Battery Monitor for High-Voltage Charging Circuits

The process of charging high-voltage batteries is a tricky one, and it's essential that the charge be closely monitored to avoid damage to the battery as well as danger to operators. This circuit uses the Burr-Brown ISO102 signal isolation buffer amplifier to isolate the monitoring from the charging circuits. Source: "Integrated Circuits Data Book Supplement," Burr-Brown.

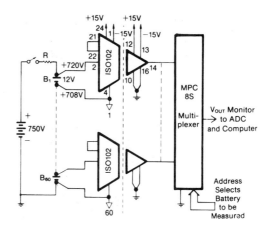

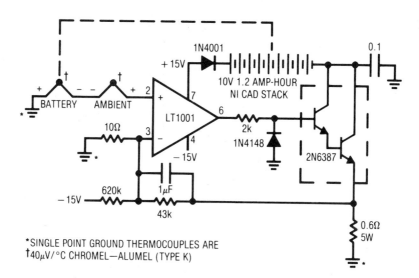

*SINGLE POINT GROUND THERMOCOUPLES ARE
†40μV/°C CHROMEL—ALUMEL (TYPE K)

Thermally Controlled NiCad Battery Charger

If you use a high current rate to charge NiCad batteries, they charge very quickly. The problem is that internal heating harms the battery and causes gas venting. This circuit solves the problem by using two thermocouples. One measures cell temperature and tapers the charge; the other nulls out the effects of ambient temperature. Source: Jim Williams, "Applications of New Precision Op Amps," Application Note 6, Linear Technology Corporation.

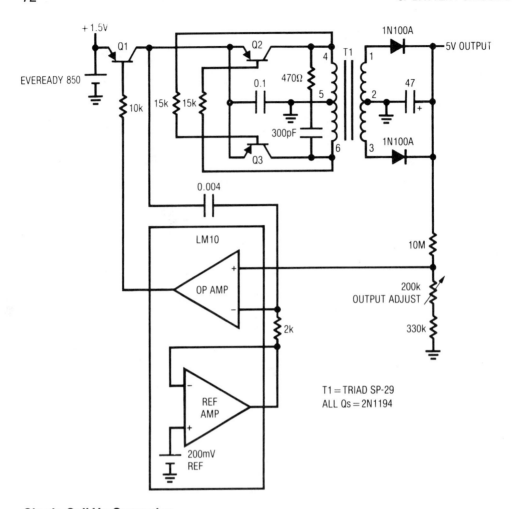

Single-Cell Up Converter

This regulated up converter uses a transformer to achieve a large voltage gain, giving you a 5-V output from a single 1.5-V cell. The circuit will supply a 5-V 150-μA load (about 25 CMOS SSI ICs) for about 3000 hours from a "D" battery. Source: Jim Williams, "Power Conditioning Techniques for Batteries," Application Note 8, Linear Technology Corporation.

5. BATTERY CIRCUITS

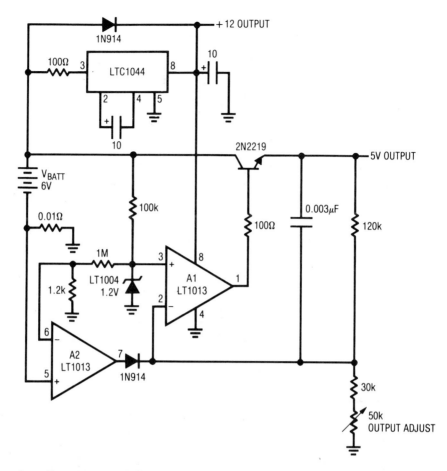

Low Dropout 5-volt Regulator
Linear regulators for batteries are a good way to get low-noise, fast transient response regulation. But to maximize battery life, it's desirable to achieve this performance with a very low regulator dropout voltage. This circuit fulfills these requirements, with a 760-μA quiescent current. In addition, its 100-mA capacity output is short circuit protected. Source: Jim Williams, "Power Conditioning Techniques for Batteries," Application Note 8, Linear Technology Corporation.

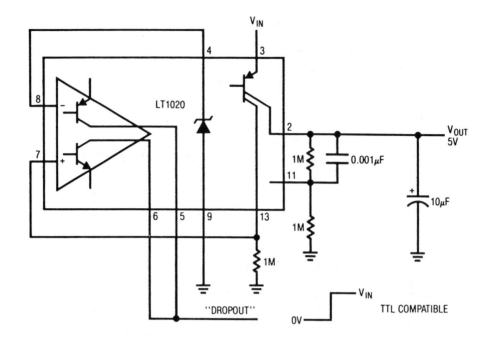

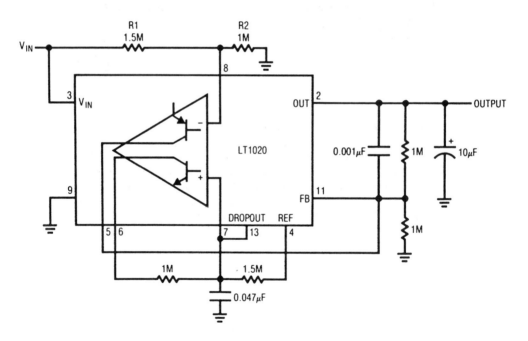

◄ Computer Battery Control

These two LT1020 micropower regulator-based circuits are useful in processor-based systems where it's desirable to monitor or control the power-down sequence. The first circuit produces a logical "1" output when the battery voltage is low, causing the regulator output to drop. You can use this circuit to alert the processor that the power is about to go down. This regulator is programmed for a 5-V output using 1-Mohm feedback resistors. The second circuit is similar, except it turns the power completely off when the dropout begins to occur. This prevents unregulated-supply conditions. It's also designed so that battery "creep back" won't cause oscillation. Source: Jim Williams, "Micropower Circuits for Signal Conditioning," Application Note 23, Linear Technology Corporation.

Chapter 6
Control Circuits

Zero-Voltage, On–Off Controller with Isolated Sensor
"Tachless" Motor Speed Controller
Wall-Type Thermostat
Low-Cost Touchtone Decoder
On–Off Touch Switch
Precision Temperature Controller
Alarm System
Overload-Protected Motor Speed Controller
Fan-Based Temperature Controller
Line-Isolated Temperature Controller
Dual-Output Over-Under Temperature Controller
Dual Time-Constant Tone Decoder
3-Phase Sine Wave Circuit
Sensitive Temperature Controller
Temperature Control
Squelch Control
Frequency Meter with Low-Cost Lamp Readout
Wide-Band Tone Detector
Motor Speed Control
Automatic Light Control

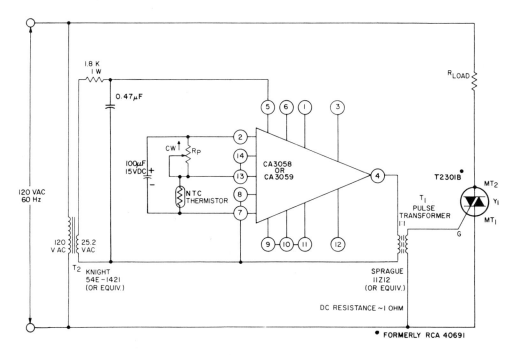

Zero-Voltage, On–Off Controller with Isolated Sensor

In many switching applications, it's necessary to isolate a sensor from the AC line. This circuit uses an RCA zero-voltage switch IC, and uses a pulse transformer to provide isolation. Source: A.C.N. Sheng, G.J. Granieri, J. Yellin, and T. McNulty, "Features and Applications of RCA Integrated Circuit Zero-Voltage Switches," Linear Integrated Circuits Monolithic Silicon, Application Note ICAN-6182, RCA Solid State Division.

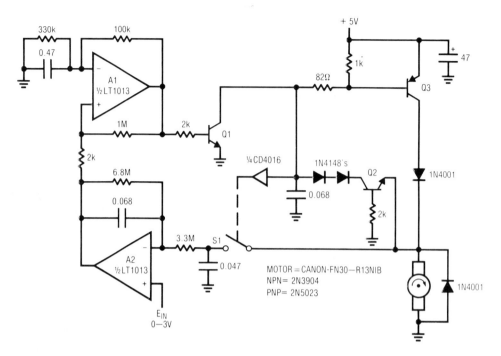

"Tachless" Motor Speed Controller

This 5-V circuit shows a way to servo control the speed of a DC motor. This motor is particularly useful in digitally controlled systems with robotic and X–Y positioning applications. By functioning from the 5-V logic supply, it eliminates additional motor drive supplies through "tachless" feedback. The circuit senses the motor's back EMF to determine its speed. The difference between the speed and a setpoint is used to close the A-sampled loop around the motor. This circuit controls from 20 rpm to full speed with good transient response under all shaft loads. Source: Jim Williams, "Designing Linear Circuits for 5V Operation," Application Note 11, Linear Technology Corporation.

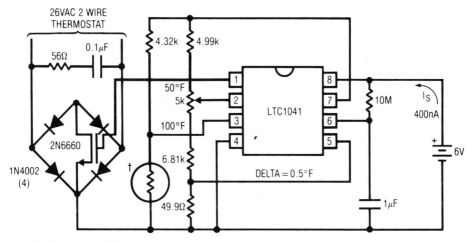

Wall-Type Thermostat

The LTC1041 dual micropower comparator is the basis for this circuit. It senses temperature using a thermistor connected in a bridge with a potentiometer (which you use to set the temperature). An important feature of this circuit is that the bridge isn't driven from the battery, but from pin 7 of the LTC1041. Pin 7 is the pulsed-power output and turns on only while the LTC1041 is sampling the inputs. In this configuration, the total system power consumption is below 1 μA. This is far less than the self-discharge rate of the battery, meaning battery life is limited only by shelf life. A typical lithium battery will run this circuit for 10–20 years. Source: Jim Williams, "Micropower Circuits for Signal Conditioning," Application Note 23, Linear Technology Corporation.

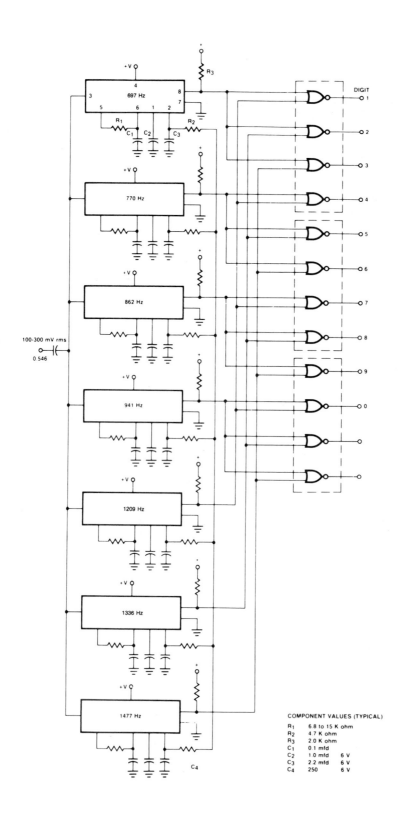

COMPONENT VALUES (TYPICAL)

R_1	6.8 to 15 K ohm	
R_2	4.7 K ohm	
R_3	2.0 K ohm	
C_1	0.1 mfd	
C_2	1.0 mfd	6 V
C_3	2.2 mfd	6 V
C_4	250	6 V

6. CONTROL CIRCUITS

◄ Low-Cost Touchtone Decoder

Because touchtone encoders (standard tone telephones) are so prevalent, touchtone decoding is useful for a wide range of control applications. This circuit uses seven XR-567 integrated dual-tone decoders, their inputs connected in common to a phone line or acoustic coupler. They drive three integrated NOR gate packages. Each tone decoder is tuned, by means of R_1 and C_1, to one of the seven tones. Source: "Dual Tone Decoding with XR-567 and XR-2567," AN-08, EXAR Databook, EXAR Corporation.

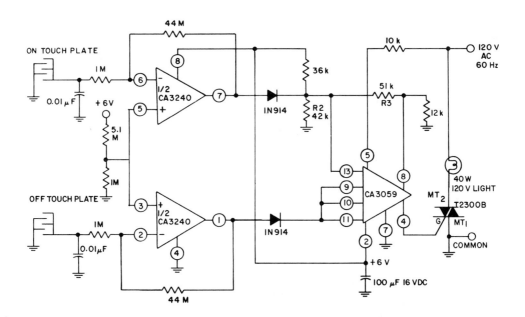

On–Off Touch Switch

This switch uses the RCA CA3240E to sense small currents flowing between two contact points on a touch plate that consists of a PC board metallization grid. When you touch the "on" plate, current flows between the two halves of the grid, causing a positive shift in the output voltage of the CA3240E. These positive transitions are fed into the CA3059, which is used as a latching circuit and zero-crossing triac driver. The advantage of using the CA3240E in this circuit is that it can sense the small currents associated with skin conduction while maintaining sufficiently high circuit impedance to protect against electrical shock. Source: A.C.N. Sheng, G.J. Granieri, J. Yellin, and T. McNulty, "Features and Applications of RCA Integrated Circuit Zero-Voltage Switches," Linear Integrated Circuits Monolithic Silicon, Application Note ICAN-6182, RCA Solid State Division.

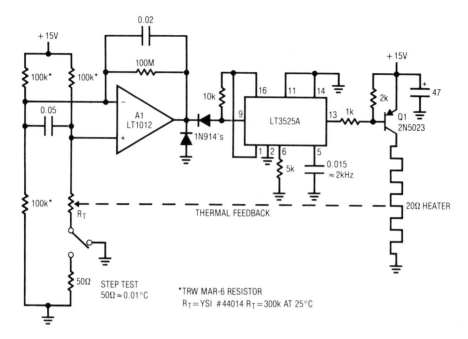

Precision Temperature Controller

This is a temperature controller for use with a small components oven. The thermistor in conjunction with the 20-ohm heater form a feedback-controlled loop. The 2-kHz pulse-width modulated heater power is much faster than the thermal loop's response and the oven sees an even, continuous heat flow. Source: Jim Williams, "Thermal Techniques in Measurement and Control Circuitry," Application Note 5, Linear Technology Corporation.

Overload-Protected Motor Speed Controller ▶

Linear Technology's LT1010 power buffer can drive difficult loads, here in a motor–tachometer combination. The tachometer signal is fed back and compared to a reference current and the 301-A amplifier closes a control loop. Because the tachometer output is bipolar, the speed is controllable in both directions with clean transitions through zero. Also, the LT1010's thermal protection prevents device destruction in the event of mechanical overload or malfunction. Source: Jim Williams, "Applications for a New Power Buffer," Application Note 4, Linear Technology Corporation.

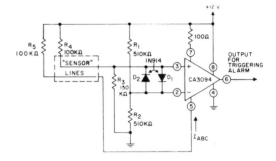

Alarm System

This simple alarm circuit uses two sensor lines. In its no-alarm state, the potential at terminal 2 is lower that the potential at terminal 3, and terminal 5 is driven with sufficient current through resistor R_5 to keep the output voltage high. If either sensor line is opened, shorted to ground, or shorted to the other sensor, the output goes low and activates whatever type of alarm you've connected to the circuit. Source: L. R. Campbell and H. A. Wittlinger, "Some Applications of a Programmable Power Switch/Amplifier," Application Note ICAN-6048, RCA Solid State.

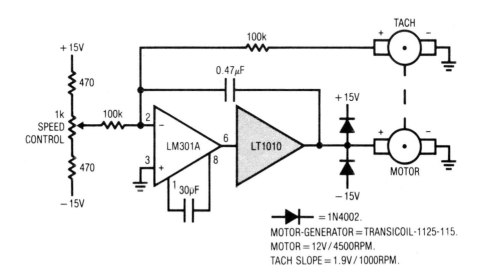

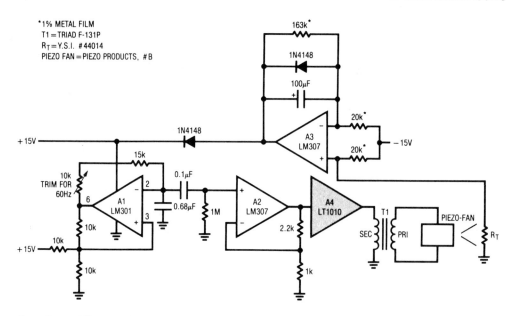

Fan-Based Temperature Controller

This circuit uses a power buffer to control a fan, which in turn regulates instrument temperature. The fan used is a electrostatic type that has very high reliability because it has no moving parts. When you apply power, the thermistor (located in the fan's exhaust stream) is at a high value and the fan doesn't run. But as the instrument enclosure warms, the thermistor value decreases and voltage is generated to run the fan. In this way, the loop acts to maintain a stable instrument temperature by controlling the fan's exhaust rate. Source: Jim Williams, "Applications for a New Power Buffer," Application Note 4, Linear Technology Corporation.

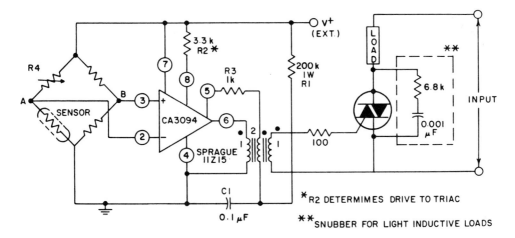

Line-Isolated Temperature Controller

Many control applications require line isolation but not zero-voltage switching (as the CA3094 is capable of). This line-isolated temperature controller is for use with inductive or resistive loads. In temperature monitoring or related control applications, the sensor can be a temperature-dependent element such as a resistor, thermistor, or diode. The load can be a lamp, bell, horn, recorder, or other appropriate device connected in a feedback relationship to the sensor. Source: A.C.N. Sheng, G.J. Granieri, J. Yellin, and T. McNulty, "Features and Applications of RCA Integrated Circuit Zero-Voltage Switches," Linear Integrated Circuits Monolithic Silicon, Application Note ICAN-6182, RCA Solid State Division.

6. CONTROL CIRCUITS

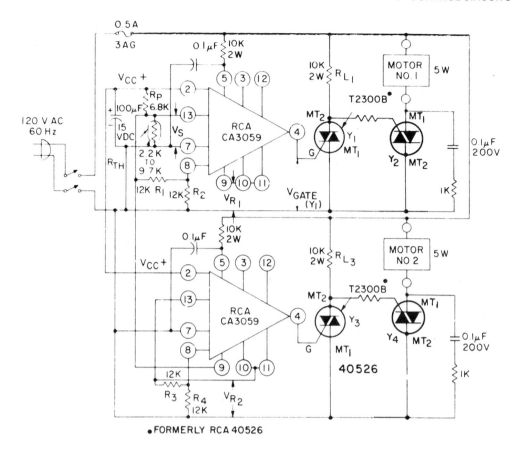

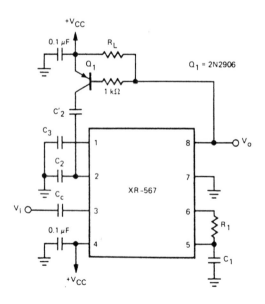

Dual Time-Constant Tone Decoder

For some applications, it's important to have a tone decoder with both narrow bandwidth and fast response time. This dual time constant decoder fulfills both these requirements. The circuit has two low-pass loop filter capacitors C_2 and C'_2. With no signal present, the input at pin 8 is high, transistor Q_1 is off, and C'_2 is switched out of the circuit. The loop low-pass filter comprises C_2, which can be kept as small as possible for minimum response time. When an in-band signal is detected, the output at pin 8 will go low, Q_1 will turn on, and capacitor C'_2 will be switched in parallel with C_2. Source: EXAR Databook, EXAR Corporation.

◂ Dual-Output Over–Under Temperature Controller

The problem of driving inductive loads such as motors by the narrow pulses generated by the RCA CA3059 zero-voltage switch is solved in this circuit by the use of the sensitive-gate RCA-40526 triac. The high sensitivity of the device (3 milliamperes maximum) and low latching current (approximately 9 milliamperes) permit synchronous control of this dual circuit. Source: A.C.N. Sheng, G.J. Granieri, J. Yellin, and T. McNulty, "Features and Applications of RCA Integrated Circuit Zero-Voltage Switches," Linear Integrated Circuits Monolithic Silicon, Application Note ICAN-6182, RCA Solid State Division.

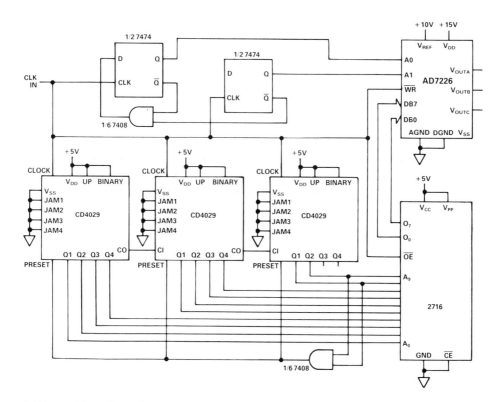

3-Phase Sine Wave Circuit

This circuit generates 3-phase sine waves that can be used for driving small 3-phase motors. It uses Analog Devices AD7226 — a quad, 8-bit, CMOS D/A converter packaged in a 20-pin DIP. The 2716 holds the sine wave in digital form. See the original application note for a BASIC program that generates the code values, as well as details on a self-programmable reference circuit that uses the AD7226's fourth — otherwise unused — D/A converter. Source: Mike Byrne, "3-Phase Sine Wave Generation Using the AD7226 Quad DAC," Application Note E924-15-7/85, Analog Devices.

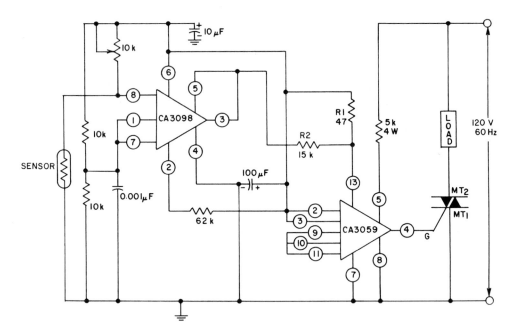

Sensitive Temperature Controller

This circuit uses an RCA CA3098E IC programmable comparator with a CA3059 zero-voltage switch. Because the CA3098E contains an integral flip-flop, its output will be either a "0" or "1" state. Consequently, the zero-voltage switch can't operate in the linear mode, and spurious half-cycling operation is prevented. Source: A.C.N. Sheng, G.J. Granieri, J. Yellin, and T. McNulty, "Features and Applications of RCA Integrated Circuit Zero-Voltage Switches," Linear Integrated Circuits Monolithic Silicon Application Note ICAN-6182, RCA Solid State Division.

6. CONTROL CIRCUITS

Temperature Control

A couple of transistors and a thermistor in the charging network of the 555-type timer enable this device to sense temperature and produce a corresponding frequency output. This circuit is accurate to ±1 Hz over a 78°F temperature range. Source: "NE555 And NE556 Applications," AN170, Linear Data Manual Volume 2: Industrial, Signetics Corporation.

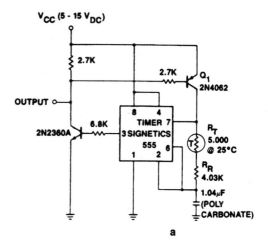

a

NOTE:
All resistor values are in ohms.

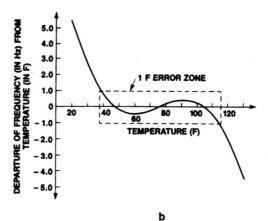

b

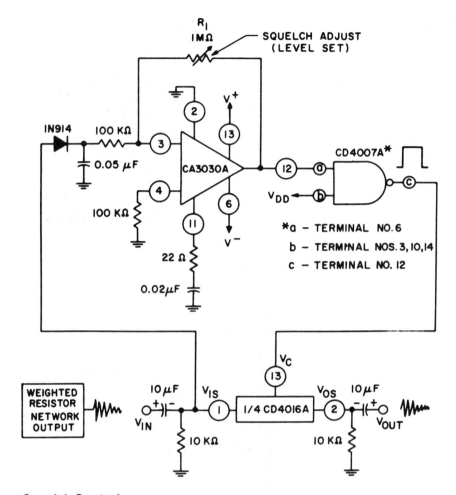

Squelch Control

Squelch circuits are used in both communication systems and in systems in which the signal must have a specific voltage level to be transmitted. In this simple squelch control, when the input signal reaches a voltage level (determined by the 1-meg resistor), the RCA CD4016A quad bilateral switch will be gated "on" and the output signal will be the same as the input. When the input signal is below this value, the CD4016A will be gated "off" and there will be no output. The variable resistance (which is normally a front-panel control) permits the adjustment of the squelch threshold. Source: J. Litus, Jr., S. Niemiec, and J. Paradise, "Transmission and Multiplexing of Analog or Digital Signals Utilizing the CD4016A Quad Bilateral Switch," Digital Integrated Circuits Application Note ICAN-6601, RCA Solid State Division.

6. CONTROL CIRCUITS

Frequency Meter with Low-Cost Lamp Readout

This circuit shows how you can connect two XR-567 integrated dual-tone decoders with overlapping detection bands for use as a go/no-go frequency meter. Unit 1 is set 6% above the desired sensing frequency and Unit 2 is set 6% below the desired frequency. If the incoming frequency is within 13% of the desired frequency, either Unit 1 or Unit 2 will give an output. If both units are on, it means that the incoming frequency is within 1% of the desired frequency. Three light bulbs and a transistor allow low-cost readout. Source: "Dual Tone Decoding with XR-567 and XR-2567," AN-08, EXAR Databook, EXAR Corporation.

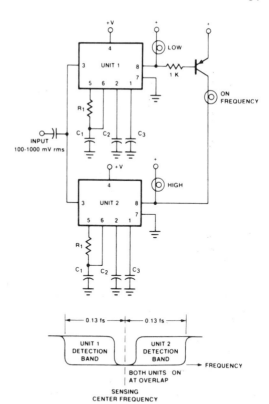

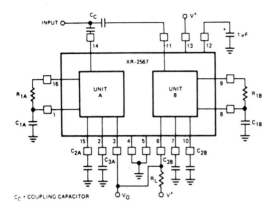

C_C = COUPLING CAPACITOR

Wide-Band Tone Detector

You can increase the detection bandwidth of the EXAR XR-2567 Dual Monolithic Tone Decoder by combining the respective bandwidths of individual decoder sections. If the detection bands of each circuit are located adjacent to each other, and if the two outputs (pins 3 and 6) are shorted together, then the resulting bandwidth is the sum of individual bandwidths. This lets you increase the total detection bandwidth to 24% of center frequency. To maintain uniform response throughout the pass band, the input signal level should be $\geq$ 80 mV RMS, and the respective passbands of each section should have approximately 3% overlap at center frequency. Source: EXAR Databook, EXAR Corporation.

6. CONTROL CIRCUITS

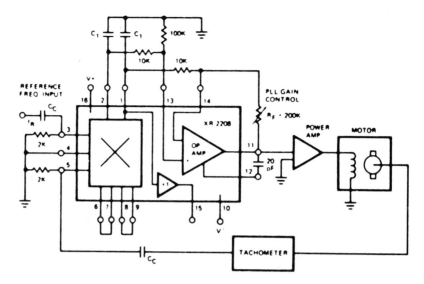

Motor Speed Control

This circuit is a motor speed control where the frequency of the motor is phase locked to the input reference frequency. The multiplier section of the EXAR XR-2208 multiplier is used as a phase comparator, comparing the phase of the tachometer output signal with the phase of the reference input. The resulting error voltage across pins 1 and 2 is low-pass filtered by capacitor C1 and amplified by the op amp section. This error signal is then applied to the motor field winding to phase-lock the motor speed to the input reference frequency. Source: EXAR Databook, EXAR Corporation.

Automatic Light Control

This control turns on a lamp when the input to the photodiode falls below a certain value. It uses the LT1083, which handles loads up to 7.5 A. Source: "Linear Databook Supplement," Linear Technology Corporation.

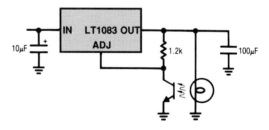

Chapter 7
Converters

1-Hz–1.25-MHz Voltage-to-Frequency Converter
1-Hz–30-MHz Voltage-to-Frequency Converter
16-bit Analog-to-Digital Converter
Cyclic Analog-to-Digital Converter
Current-to-Voltage Converter
Offset Stabilized Voltage-to-Frequency Converter
Current Loop Receiver/Transmitter
Thermocouple-to-Frequency Converter
9-bit Digital-to-Analog Converter
TTL-to-MOS Converter
Temperature-to-Frequency Converter
10-bit 100-μA Analog-to-Digital Converter
Micropower 12-Bit 300-μsec A/D Converter
Fully-Isolated 10-Bit A/D Converter
V/F–F/V Data Transmission Circuit
Quartz-Stabilized Voltage-to-Frequency Converter
2.5-MHz Fast-Response V/F Converter
Push–Pull Transformer-Coupled Converter
Ultra-High Speed 1-Hz to 100-MHz V/F Converter
Low-Power 10-kHz V/F Converter
ECL-to-TTL and TTL-to-ECL Converters
Low-Power 10-Bit A/D Converter
Basic 12-bit 12-μsec Successive Approximation A/D Converter
Wide-Range Precision PLL F/V Converter
Centigrade-to-Frequency Converter
Micropower 10-kHz V/F Converter
Micropower 1-MHz V/F Converter
RMS-to-DC Converter

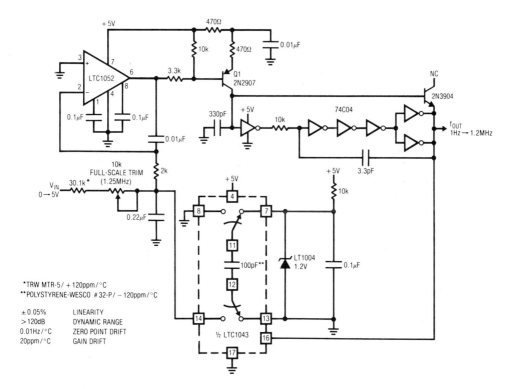

1-Hz–1.25-MHz Voltage-to-Frequency Converter

This relatively inexpensive V/F converter runs off a single 5-V battery. It has a wide 1-Hz to 1.25-MHz range, 0.05% linearity, and a temperature coefficient of typically 20ppm/°C. To avoid board leakage that can cause jitter, you should either use a clean board or mount the capacitor, Q1's collector, the inverter input, and the transistor base connection on a Teflon standoff, using short connections. To trim this circuit, apply +5.000 V and adjust the 1.25-MHz trim for 1.2500 MHz out. Source: Jim Williams, "Application Considerations and Circuits for a New Chopper-Stabilized Op Amp," Application Note 9, Linear Technology Corporation.

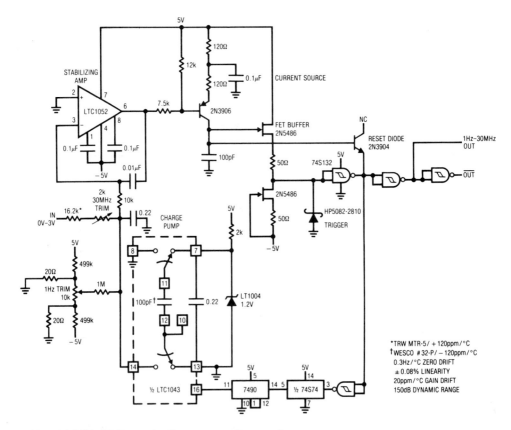

1-Hz–30-MHz Voltage-to-Frequency Converter

This circuit, with its impressive 1-Hz to 30-MHz range for a 0- to 5-V input, fully uses the 150-dB dynamic range of Linear Technology's LTC1052 chopper-stabilized amplifier. The circuit maintains 0.08% linearity over its entire 7 1/3 decade range with a full-scale drift of about 20ppm/°C. To trim this circuit, ground the input and adjust the 1-Hz trim until oscillation just starts. Next apply 5.000 V and set the 30-MHz trim for a 30.00-MHz output. Repeat this procedure until both points are fixed. Source: Jim Williams, "Application Considerations and Circuits for a New Chopper-Stabilized Op Amp," Application Note 9, Linear Technology Corporation.

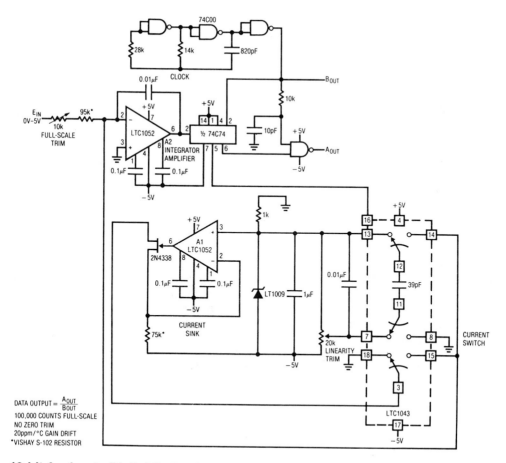

16-bit Analog-to-Digital Converter

This A/D converter uses the high-performance characteristics of Linear Technology's LTC1052 chopper-stabilized amplifier. It has an overrange to 100,000 counts and uses a current-balancing technique. To calibrate this circuit, apply 5.00000 V and adjust the full-scale trim for 100,000 counts out. Next, set the input to 1.25000 V and adjust the linearity trim for 25,000 counts out. Repeat this procedure until both points are fixed. Source: Jim Williams, "Application Considerations and Circuits for a New Chopper-Stabilized Op Amp," Application Note 9, Linear Technology Corporation.

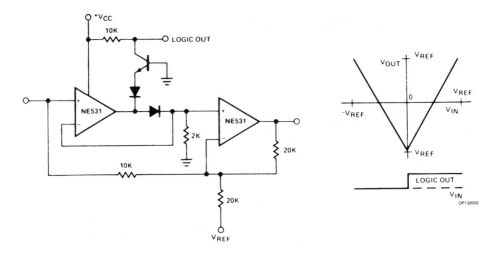

Cyclic Analog-to-Digital Converter

One interesting, but much ignored, A/D converter is the cyclic type shown here. It consists of a chain of identical stages, each of which senses the polarity of the input. The stage then subtracts V_{ref} from the input and doubles the remainder if the polarity was correct. In this circuit, the signal is full-wave rectified and the remainder of $V_{in} - V_{ref}$ is doubled. A chain of these stages gives the gray code equivalent of the input voltage in digitized form related to the magnitude of V_{ref}. Possessing high potential accuracy, this circuit settles in 5-μsec. Source: Linear Data Manual, Volume 2: Industrial, Signetics Corporation.

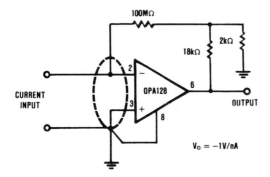

Current-to-Voltage Converter

This simple current-to-voltage converter uses the Burr-Brown OPA128 ultra-low bias current monolithic operational amplifier. Because of its advanced geometry dielectrically isolated FET (DIFET) inputs, this circuit achieves a performance exceeding even the best hybrid electrometer amplifiers. Source: "Integrated Circuits Data Book," Burr-Brown.

7. CONVERTERS

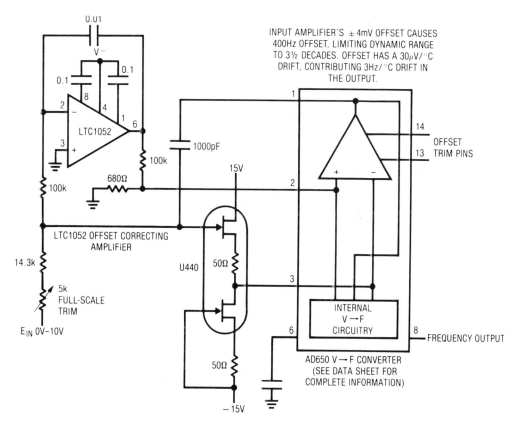

Offset Stabilized Voltage-to-Frequency Converter

This circuit shows a way to offset-stabilize a data converter, thereby doubling its dynamic range of operation. This eliminates the necessity for an offset trim and reduces offset trim to negligible levels. To calibrate, apply 10 V and trim the output for exactly 1 MHz. Source: Jim Williams, "Application Considerations and Circuits for a New Chopper-Stabilized Op Amp," Application Note 9, Linear Technology Corporation.

7. CONVERTERS

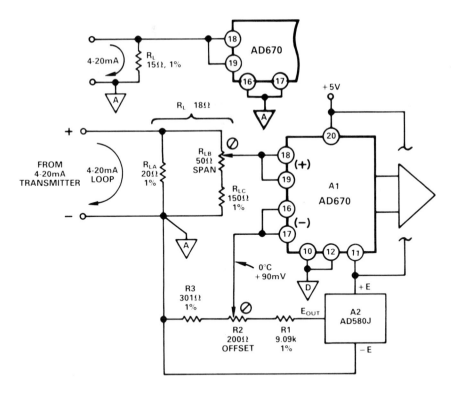

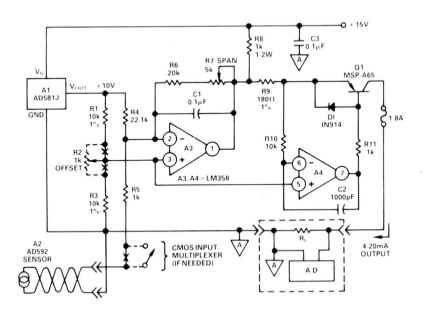

◄ Current Loop Receiver/Transmitter

These two circuits illustrate an AD670 A/D converter interfaced to a remote 4- to 20-mA loop. Since the load resistance at the receiver is low, it can accommodate line lengths of up to several thousand feet of #22 wire. The receiver to the A/D circuit receives an input signal in current form and converts it into a proportional voltage across the load resistance R_L. In this application, the transmitted signal is from an AD592 temperature sensor connected to the transmitter circuit. The transmitter is powered at a remote site from a +15-V source. If you're using multiplexing as shown, the transmitter can accept more than one temperature sensor. Calibration can be done at the A/D or at the transmitter, with R2 used for zero scale and R7 for SPAN. Source: Walt Jung, "AD670 8-Bit A/D Converter Applications," Application Note E935-12-8/85, Analog Devices.

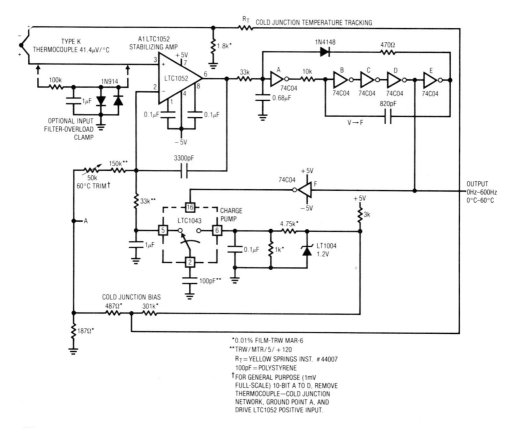

Thermocouple-to-Frequency Converter

This converter uses the popular type K thermocouple as a sensor. Its accuracy is ±1°C over a 0–60° range; and its resolution is 0.1°C. The thermocouple's characteristics, combined with A1's low offset and the cold junction/offsetting network components, eliminate zero trimming. Calibration is accomplished by placing the thermocouple in a 60°C environment and adjusting the 50K-ohm potentiometer for a 600-Hz output. Source: Jim Williams, "Some Techniques for Direct Digitization of Transducer Outputs," Application Note 7, Linear Technology Corporation.

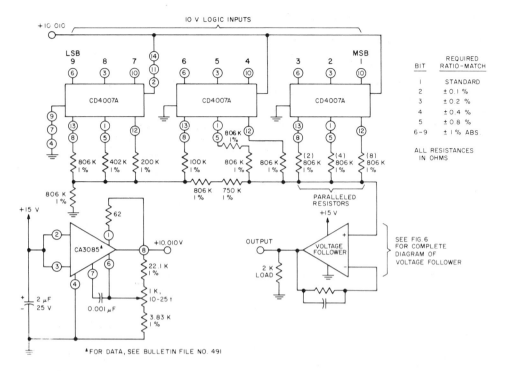

9-bit Digital-to-Analog Converter

This DAC uses three RCA CD4007A dual complementary pairs plus inverter to perform the switch function using a 10-volt logic level. A single 15 volt supply provides a positive bus for the follower amplifier and feeds the CA3085 voltage regulator. The line-voltage of approximately 0.2% permits 9-bit accuracy to be maintained with a variation of several volts in the supply. System power consumption ranges between 70 and 200 mW. The resistor ladder is composed of 1% tolerance metal-oxide film resistors. Source: O. H. Schade Jr., "Digital-to-Analog Conversion Using the RCA-CD4007A COS/MOS IC," Application Note ICAN-6080, RCA Solid State.

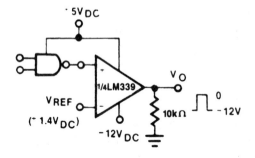

TTL-to-MOS Converter

This simple logic converter takes advantage of the LM339 voltage comparator's ability to interface to both TTL and CMOS. Source: Linear Data Manual Volume 2: Industrial, Signetics Corporation.

7. CONVERTERS

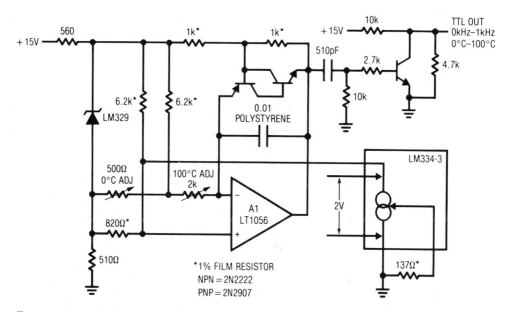

Temperature-to-Frequency Converter

Here's a simple way to convert the current output of an LM334 temperature sensor to a corresponding output frequency. To calibrate this circuit, place the LM334 in a 0°C environment and trim the 0C adjust for 0 Hz. Next, put the LM334 in a 100°C environment and set the 100C adjust for 1-KHz output. Repeat this procedure until both points are fixed. This circuit has a stable 0.1°C resolution with a ±1.0°C accuracy. Source: Jim Williams, "Some Techniques for Direct Digitization of Transducer Outputs," Application Note 7, Linear Technology Corporation.

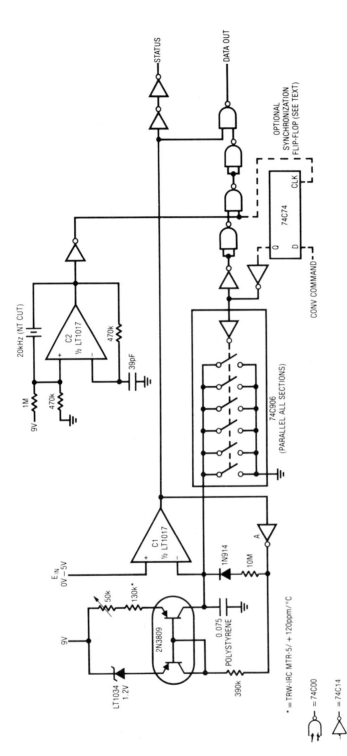

10-bit 100-µA Analog-to-Digital Converter

While this analog-to-converter has less resolution than many designs, it only requires 100µA. This circuit will typically hold ±1 accuracy over a temperature range of 0–70°C with an additional ±1 LSB because of the synchronous relationship between the clock and the conversion sequence. If the conversion sequence is synchronized to the clock, the ±1 LSB is removed, and the total error falls to ±1 LSB. The flip-flop that's shown in dashed lines permits this synchronization. This circuit's conversion rate varies with input. At tenth scale 150 Hz is possible, decreasing to 20 Hz at full-scale. Source: Jim Williams, "Micropower Circuits for Signal Conditioning," Application Note 23, Linear Technology Corporation.

7. CONVERTERS

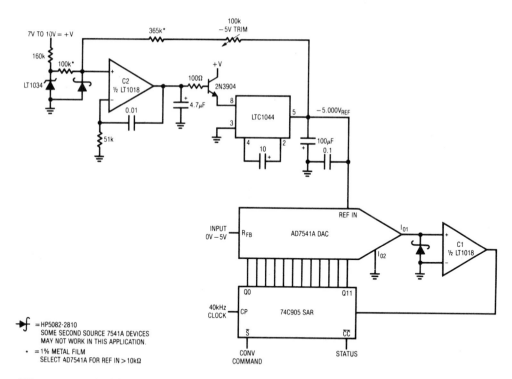

Micropower 12-bit 300-μsec Analog-to-Digital Converter

Although low-power 12-bit analog-to-digital converters are generally available, they're usually quite slow — in the 100msec range. Higher speeds require a successive approximation (SAR) approach. This circuit converts in 300 μs, while consuming only 890 μA. The typical temperature coefficient of this circuit is 30/ppm/°C. and an accuracy of ±2 LSBs. To trim this circuit, adjust the 100K potentiometer for exactly -5 V at V_{ref}. The DAC's internal feedback resistor serves as the input. Source: Jim Williams, "Micropower Circuits for Signal Conditioning," Application Note 23, Linear Technology Corporation.

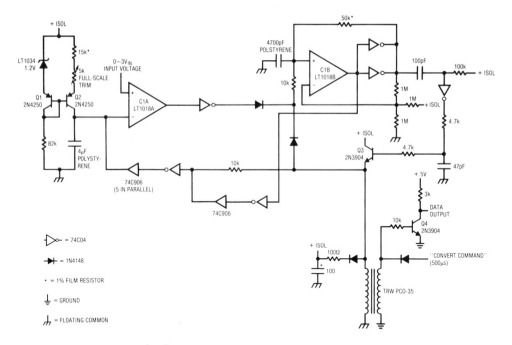

Fully Isolated 10-bit A/D Converter

This 5-V circuit is a complete 10-bit A/D converter that is fully floating from system ground. It's ideal for performing 10-bit A/D conversions in the high common-mode noise characteristics that are prevalent in predominantly-digital systems. It's also useful in industrial environments, where nosie and high common-mode voltages are present in transducer-fed systems. Circuit operation is initiated by applying a pulse to the "convert command" input. Source: Jim Williams, "Designing Linear Circuits for 5V Operation," Application Note 11, Linear Technology Corporation.

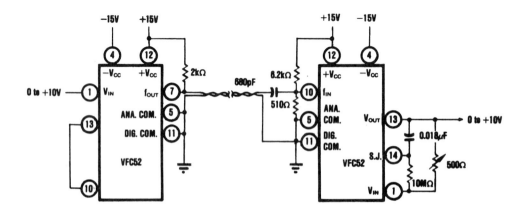

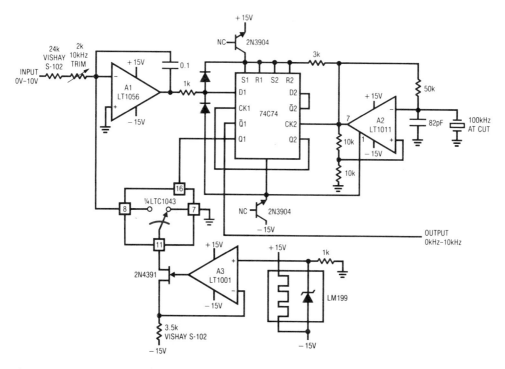

Quartz-Stabilized Voltage-to-Frequency Converter

For the lowest gain drift in a V/F converter, you need a quartz-stabilized clock. This circuit uses half of a flip-flop to produce a 50-kHz clock driven by a quartz-stabilized relaxation oscillator. This reduces drift to about 5ppm/C. The quartz crystal contributes about 0.5ppm/C, with the remaining drift a function of the current source components, switching time variations, and the input resistor. A voltage-to-frequency converter of this type is usually restricted to relatively low full-scale frequencies (e.g., 10 kHz to 100 kHz) because of speed limitations in accurately switching the current sink. To trim this circuit, apply exactly 10 V in and adjust the 2k potentiometer for 10.000-kHz output. Source: Jim Williams, "Designs for High Performance Voltage-to-Frequency Converters," Application Note 14, Linear Technology Corporation.

◀ V/F–F/V Data Transmission Circuit

A novel use for a voltage-to-frequency and frequency-to-voltage converters is to convert analog data into a digital pulse train for transmission over long lines through high EMI environments. As shown here, you can use the combination to transmit analog data of 0 to +10V over a 100-ohm shielded twisted pair. Source: "Integrated Circuits Data Book," Burr-Brown.

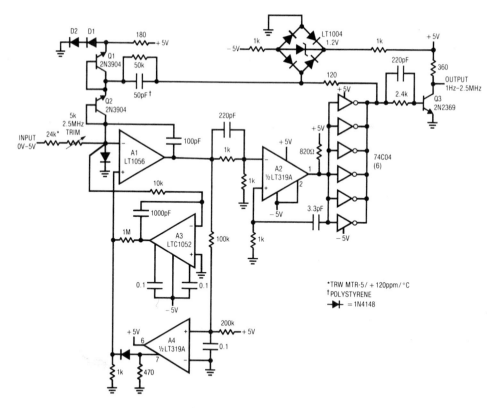

2.5-MHz Fast-Response V/F Converter

With a 2.5 MHz output that settles from a full-scale input step in only 3 μsec, this circuit makes a good candidate for FM applications or any area where you need fast response to input movement. Its speed results from the lack of a servo amplifier. Instead the charge is fed back directly to the oscillator for immediate response. To trim the circuit, apply 5.000 V and adjust the 5k potentiometer for a 2.500-MHz output. A3's low offset eliminates the requirement for a zero trim. Source: Jim Williams, "Designs for High Performance Voltage-to-Frequency Converters," Application Note 14, Linear Technology Corporation.

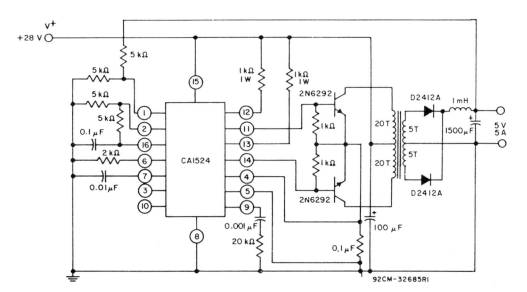

Push–Pull Transformer-Coupled Converter

The output stage of the RCA CA1524 pulse width modulator IC provides the drive for transistors Q1 and Q2 in the push–pull application shown here. Since the internal flip–flop drives the oscillator frequency by two, the oscillator must be set at twice the output frequency. Current limiting for this circuit is done in the primary of transformer T1 so that the pulse width will be reduced if transformer saturation occurs. Source: Carmine Salerno, "Application of the CA1524 Series Pulse-Width Modulator ICs," Linear Integrated Circuits Application Note ICAN-6915, RCA Solid State Division.

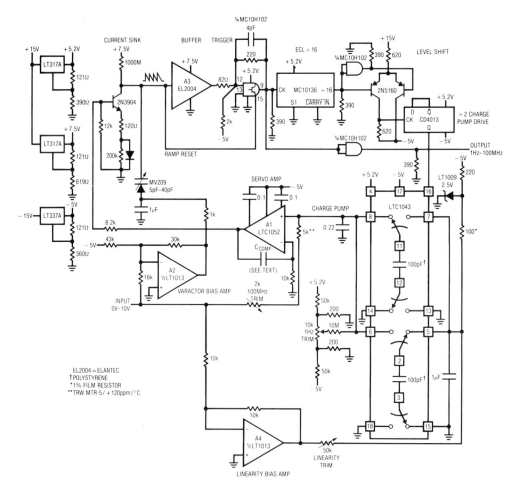

Ultra-High-Speed 1-Hz to 100-MHz V/F Converter

This circuit uses a variety of methods to achieve wider dynamic range (160 dB) and higher speed than most commercial voltage-to-frequency converters. The design provides 10% overrange to 110 MHz. The circuit characteristics are derived from the basic V/F characteristics. The chopper-stabilized amplifier and charge pump stabilize the circuit's operating point, contributing to high linearity and low drift. To calibrate this circuit, apply 10.000V and trim the 100-MHz adjustment for 100.000 MHz. If you don't have a fast enough counter, the /32 signal at pin 16 of the LTC1043 will read 3.1250 MHz. Next, ground the input, install C_{comp} = 1µF and adjust the 1-Hz trim until the circuit oscillates at 1 Hz. Finally, set the linearity trim for 50.000 MHz for a 5.000 V input. Repeat these adjustments until all three points are fixed. Source: Jim Williams, "Designs for High Performance Voltage-to-Frequency Converters," Application Note 14, Linear Technology Corporation.

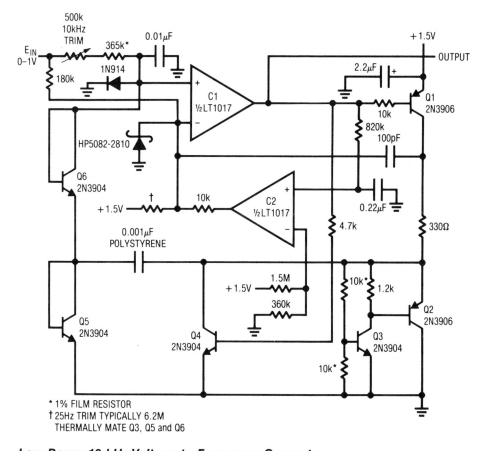

Low-Power 10-kHz Voltage-to-Frequency Converter

Although circuits powered by a single-cell 1.5-V battery are desirable for a wide range of applications, the low voltage eliminates most linear ICs as design candidates. As a further complication, these circuits need to function at end-of-life battery voltage. This circuit is a complete 1.5-V powered 10-kHz voltage-to-frequency converter. A 0- to 1-V input produces a 25-Hz to 10-kHz output, with a transfer linearity of 0.35%. Gain drift is 250ppm/C and current consumption about 800 μA. To calibrate this circuit, apply 2.5 mV at the input and select the resistor value at C1's input for a 25-Hz output. Then put in exactly 1 V and trim the 500k potentiometer for 10-kHz output. Source: Jim Williams, "Circuitry for Single Cell Operation," Application Note 15, Linear Technology Corporation.

ECL-to-TTL and TTL-to-ECL Converters

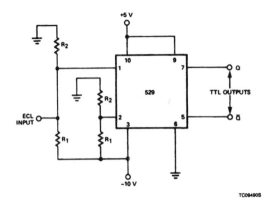
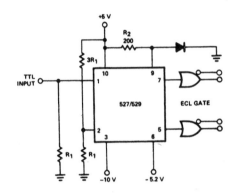

Emitter-coupled logic is popular because of its speed, and many circuits are built with the higher-speed portions in ECL. These circuits solve the problem of interfacing the two ECL and TTL. The first circuit converts ECL to TTL. The standard logic output swings of ECL are -0.8 V to 1.8 V at room temperature. Converting these signals to TTL levels is accomplished simply by using the basic voltage comparator circuit with slight modifications. As shown here, the power supplies have been shifted in order to shift the common-mode range more negative. This insures that the common-mode range isn't exceeded by the logic inputs. Since ECL is inherently fast, the Signetics NE529 is a good choice because of its speed. The second circuit, operating in reverse, again uses the NE529 to convert TTL to ECL. Source: "Applications for the NE521/522/527/529," AN116, Linear Data Manual Volume 2: Industrial, Signetics Corp.

7. CONVERTERS

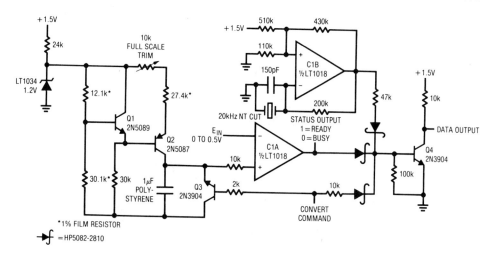

Low-Power 10-bit A/D Converter

This integrating analog-to-digital converter has a 60-msec conversion time. It consumes 360-μA from its 1.5-V supply and maintains 10-bit accuracy over a 15- to 35°C temperature range. The number of pulses appearing at the output is directly proportional to the input voltage. To calibrate this circuit, apply 0.5000 V to the input and trim the 10k potentiometer for exactly 1000 pulses out each time the convert command line is pulsed. Source: Jim Williams, "Circuitry for Single Cell Operation," Application Note 15, Linear Technology Corporation.

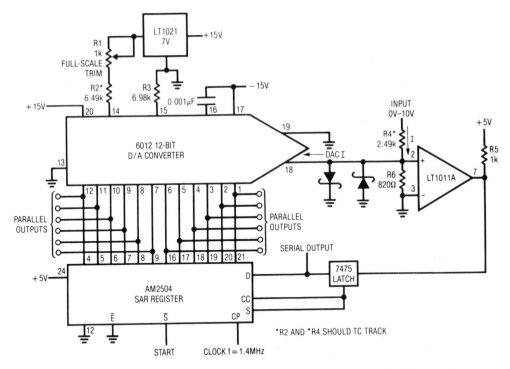

Basic 12-bit 12-μsec Successive Approximation Analog-to-Digital Converter

The most popular A/D method used today is the successive approximation register (SAR) converter. At the 12-bit level, most currently available monolithic devices require about 10 μsec to convert. And even though modular or hybrid units achieve conversion speeds below 2 μsec, they're usually quite expensive. Because of these factors, it's often more desirable to build your own 12-bit SAR converter, and this simple circuit will fill the bill, converting in about 12 μsec. Source: Jim Williams, "Considerations for Successive Approximation A-D Converters," Application Note 17, Linear Technology Corporation.

7. CONVERTERS

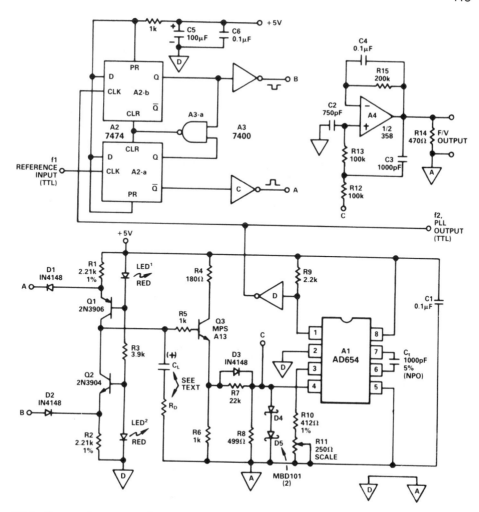

Wide-Range Precision PLL Frequency-to-Voltage Converter

Since discrete phase locked loops (PLLs) are readily available, why does this circuit use an F/V converter to construct one? There are two reasons: to achieve a greater dynamic range and to implement a high-precision F/V circuit. This circuit uses a wide dynamic range digital frequency/phase detector as the PLL comparator. With this type of circuit, the input signal can range over a very wide span and the locking characteristic isn't harmonically sensitive. This circuit accepts a TTL compatible square or pulse input. To trim this circuit, adjust R11 for exactly 0.5 V at the V/F input. Source: Walt Jung, "Operation and Applications of the AD654 IC V-to-F Converter," Application Note E923-25-7/85, Analog Devices.

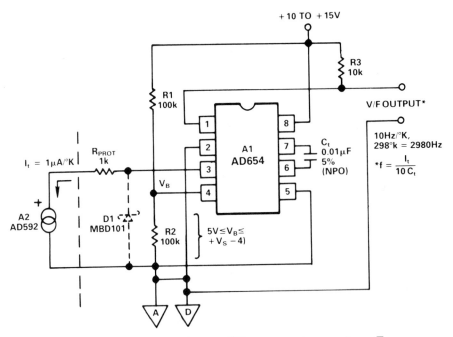

a. Kelvin-Scaled Temperature – to – F

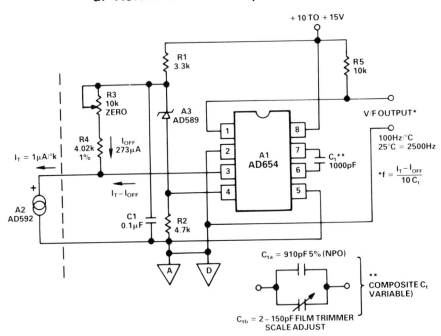

b. Centigrade-Scaled Temperature – to – F

7. CONVERTERS

◄ Centigrade-to-Frequency Converter

Today's IC temperature transducers are both low in cost and easy to use in remote measurement applications. This circuit uses an AD592 temperature sensor along with an AD654 voltage-to-frequency converter. Above 0°C, this circuit has a sensitivity of 100 Hz/°C. An interesting feature of this circuit is that the offset current is regulated by the floating reference diode A3, even though the common-mode input to the AD654 can vary with supply voltage. Trim this circuit by placing sensor A2 in an ice bath and adjust R3 until the output frequency just stops. Then place A2 in boiling water (100°C) and trim C_t for 10,000 Hz. Since the trimmer is floating, you should use a plastic tuning wand. Source: Walt Jung, "Operation and Applications of the AD654 IC V-to-F Converter," Application Note E923-25-7/85, Analog Devices.

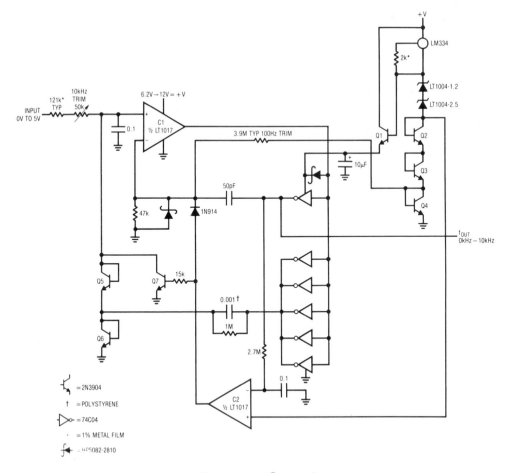

Micropower 10-kHz Voltage-to-Frequency Converter

With a maximum current consumption of only 145 μsec, this circuit's power requirements are far below most other currently available voltage-to-frequency converters. A 0- −5-V input produces a 0- −10-kHz output, with a linearity of 0.02%. To calibrate this circuit, apply 50 mV and select the value at C1's input for a 100 Hz output. Then apply 5 V and trim the input potentiometer for a 10-kHz output. Source: Jim Williams, "Micropower Circuits for Signal Conditioning," Application Note 23, Linear Technology Corporation.

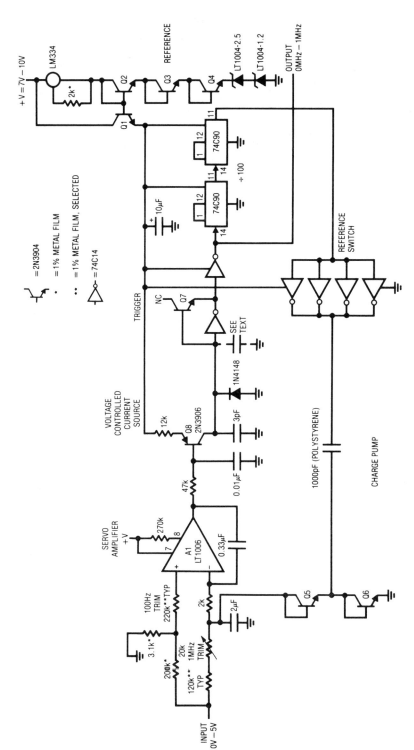

Micropower 1-MHz Voltage-to-Frequency Converter

This circuit runs at 1 MHz full-scale. Its quiescent current is 245 uA, ascending linearly to 635 μA at 1-MHz output. To trim this circuit, put in 500 μV and select the indicated value at A1's positive input for 100 Hz out. Then put in 5 V and trim the 200-K potentiometer for 1 MHz out. Repeat the procedure until both points are fixed. Source: Jim Williams, "Micropower Circuits for Signal Conditioning," Application Note 23, Linear Technology Corporation.

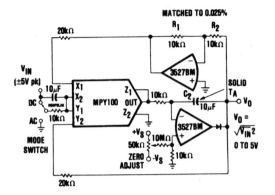

RMS-to-DC Converter

This true RMS-to-DC conversion circuit is built around an MPY100 multiplier-divider. It gives greater accuracy and bandwidth but with less dynamic range than most RMS-to-DC converters. Source: "Integrated Circuits Data Book," Burr-Brown.

Chapter 8

Filter Circuits

Low-Pass Filter
Loop Filter
Voltage-Controlled, Second-Order Filter
Voltage-Controlled Filters
Tracking Filter
Digitally Tuned Switched Capacitor Filter
10-Hz, Fourth-Order Butterworth Low-Pass Filter
Clock-Tunable Notch Filter
Bi-Quad Filter
Voltage-Controlled Circuits
Multi-Cutoff-Frequency Filter
Lowpass Filter with 60-Hz Notch
60-Hz Reject Filter
Voltage-Controlled Filter

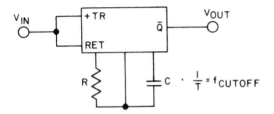

Low-Pass Filter

This very simple circuit uses the RCA CD4047A monostable/astable multivibrator. The time constant you choose for the multivibrator determines the upper cutoff frequency for the filter. This circuit essentially compares the input frequency with its own reference, and produces an output that follows the input for frequencies less than f_{cutoff}. Source: J. Paradise, "Using the CD4047A in COS/MOS Timing Applications," Digital Integrated Circuits Application Note ICAN-6230, RCA Solid State Division.

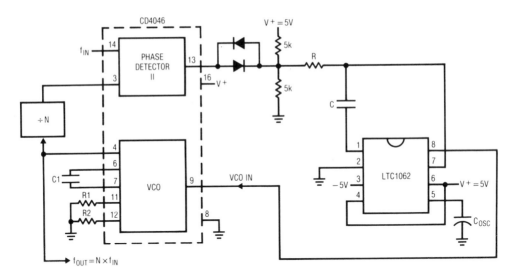

Loop Filter

With commercially available PLLs, you can design loop filters to optimize loop performance. For many applications, a first or second-order lowpass passive or active R,C filter will do the job. But when you need both minimum output jitter and good transient response at the same time, the loop filter design becomes more sophisticated. For instance, you need wide filter bandwidth for good transient response, as well as minimum jitter at VCO input for minimum jitter at the output. In this circuit, the LTC1062 lowpass filter fills the need, while providing economy and programmable cutoff frequency. Source: Nello Sevastopoulos and Phillip Karantzalis, "Unique Applications for the LTC1062 Lowpass Filter," Application Note 24, Linear Technology Corporation.

8. FILTER CIRCUITS

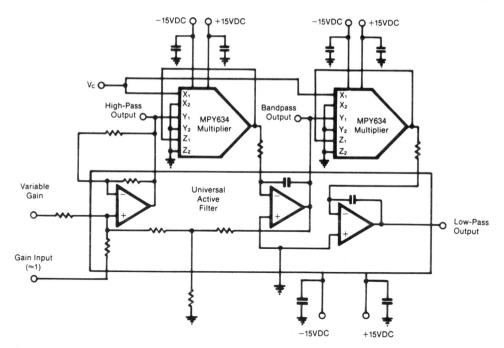

Voltage-Controlled, Second-Order Filter

This circuit is designed around two Burr-Brown multiplier chips and a UAF21 universal active filter. The multipliers act as linear voltage-controlled resistors that vary the center and cutoff frequencies of the active filter chip. Using just a handful of components, this design gives you bandwidths up to 200 kHz and Q values up to 500. Source: "The Handbook of Linear IC Applications," Burr-Brown.

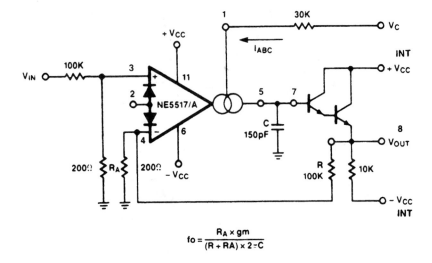

$$f_o = \frac{R_A \times g_m}{(R + R_A) \times 2 \pi C}$$

TC21900S

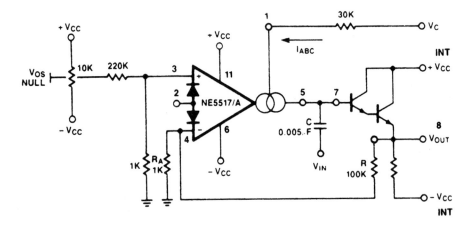

Voltage-Controlled Filters

These two filters use the Signetics NE5517A dual operational transconductance amplifier. The first is a low-pass filter. Below the corner frequency the circuit has an amplification of 0 dB. Above the corner frequency the attenuation drops by 6 dB/octave. The second circuit is a high-pass filter that's built in a similar manner, except the input is coupled via capacitor. Source: Linear Data Manual Volume 2: Industrial, Signetics Corporation.

8. FILTER CIRCUITS

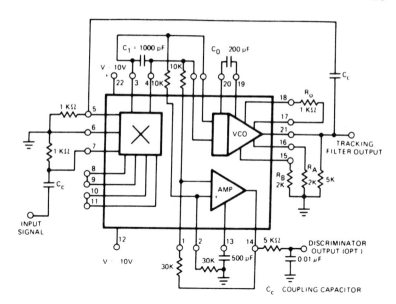

Tracking Filter

This filter tracks the input over a broad range of frequencies around the VCO cutoff frequency. It can track input signals over a 3:1 frequency range. Based around EXAR's XR-S200 multi-function PLL system chip, when the PLL locks on an input signal, it produces a filtered version of the input signal frequency at the VCO output. Source: EXAR Databook, EXAR Corporation.

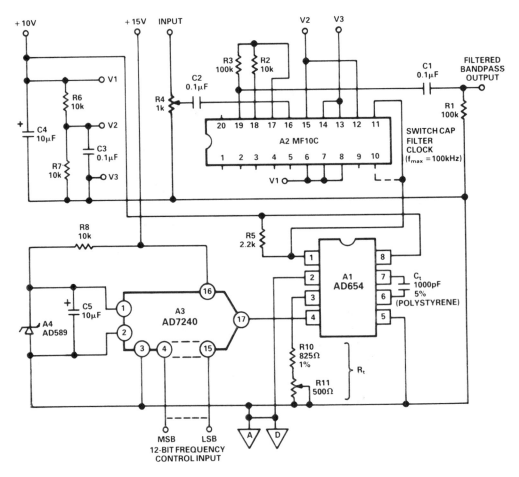

Digitally Tuned Switched-Capacitor Filter

Often called "switchcaps," IC-based switched-capacitor filters are popular because of the all around utility of the state variable filter itself. It allows you to set all standard filter modes using either resistors or pin programming. You can also use a clock input to easily and precisely tune the filter center frequency. In this circuit, the output of an AD654 voltage-to-frequency converter is used as a variable-rate clock for driving the switchcap. The filter's center frequency is programmable over a dynamic range of 2^{12}. When this circuit is calibrated for an actual filter frequency of 1 kHz with an all 1's D/A input, its accuracy will be ±1.5% or better. Source: Walt Jung, "Operation and Applications of the AD654 IC V-to-F Converter," Application Note E923-25-7/85, Analog Devices.

10-Hz, Fourth-Order Butterworth Low-Pass Filter

This circuit uses Burr Brown's OPA2111, a high-precision, monolithic DIFET (dielectrically isolated FET) operational amplifier. As in any situation where high impedances are involved, you should carefully shield the input to reduce hum pickup. Source: "Integrated Circuits Data Book," Burr-Brown.

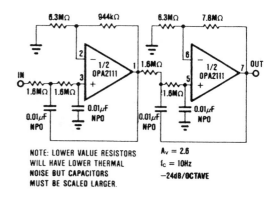

NOTE: LOWER VALUE RESISTORS WILL HAVE LOWER THERMAL NOISE BUT CAPACITORS MUST BE SCALED LARGER.

$A_v = 2.6$
$f_c = 10Hz$
$-24dB/OCTAVE$

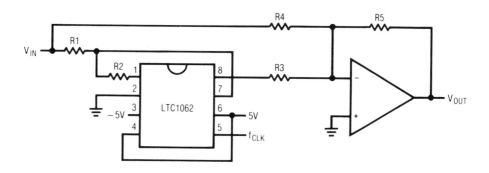

For simplicity use R3 = R4 = R5 = 10k;
$$\frac{R1}{R2} = 1.234, \frac{f_{CLK}}{f_{notch}} = \frac{79.3}{1}$$

Clock-Tunable Notch Filter

If you use a resistor between pins 1 and 7 of the LTC1062 lowpass filter, the circuit loses its lowpass characteristics and the response of the filter becomes selective like a bandpass. Also, since the two external components (R1 and R2) are frequency independent, the LTC1062 can be fully swept with an external clock. The resistor ratio mainly alters the filter's peak gain, and has very little effect on the value of the peak frequency. Because of this, you can stagger-tune two LTC1062s with a common clock to produce a respectable bandpass response. This circuit's clock-to-notch frequency ratio is a repeatable 79.3:1. Source: Nello Sevastopoulos and Phillip Karantzalis, "Unique Applications for the LTC1062 Lowpass Filter," Application Note 24, Linear Technology Corporation.

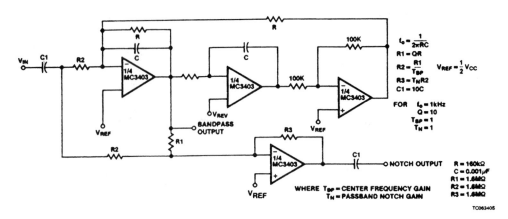

Bi-Quad Filter

This filter has both a bandpass and a notch output, and uses the Signetics MC3403 quad op amp with differential inputs. Source: "Applications for the MC3402," AN160, Linear Data Manual Volume 2: Industrial, Signetics Corporation.

8. FILTER CIRCUITS

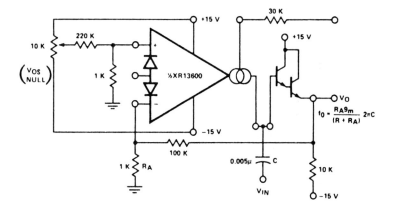

Voltage Controlled High-Pass Filter

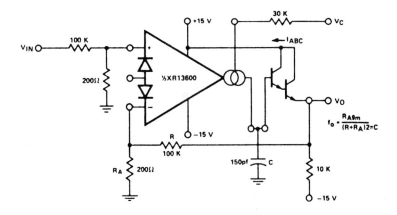

Voltage Controlled Low-Pass Filter

Voltage-Controlled Circuits

These two circuits show how EXAR's XR-13600 dual operational transconductance amplifiers are useful for designing voltage-controlled filters. There's also an added advantage in that the required buffers are included on the IC. The low-pass filter performs as a unity-gain buffer amplifier at frequencies below cut-off. At frequencies above cut-off, the circuit provides a single RC rolloff (6 dB per octave) of the input signal. The high-pass VC filter performs in much the same way, with a single RC rolloff below the defined cut-off frequency. Source: EXAR Databook, EXAR Corporation.

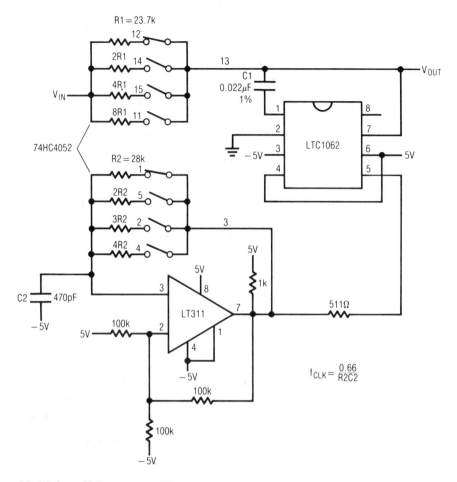

Multi-Cutoff-Frequency Filter

The LTC1062 lowpass filter lets you obtain several cutoff frequencies at the same time by simultaneously varying the clock frequency and the external R times C product. With a dual four-channel multiplexer, you can easily derive four different cutoff frequencies by selecting four input resistors and four clock frequencies. This circuit has cutoff frequencies of 500, 250, 125, and 62.5 Hz. Source: Nello Sevastopoulos and Phillip Karantzalis, "Unique Applications for the LTC1062 Lowpass Filter," Application Note 24, Linear Technology Corporation.

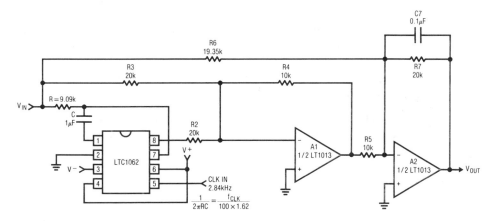

Lowpass Filter with 60-Hz Notch

Filters with a notch are generally difficult to design and require tuning. Although you can use universal switched-capacitor filters to make very precise notches, you need to limit their useful bandwidth well below half the clock frequency; otherwise, aliasing will severely limit the filter's dynamic range. You can use the LTC1062 lowpass filter to create a notch because the frequency where it exhibits a 180° phase shift is inside a passband that's repeatable and predictable from part to part. In this circuit, the input signal is summed with the lowpass filter through A1. Then the output of A1 is again summed with the input voltage A2. Source: Nello Sevastopoulos, "Application Considerations for an Instrumentation Lowpass Filter," Application Note 20, Linear Technology Corporation.

60-Hz Reject Filter

Eliminating 60-Hz AC power line hum is a common problem in numerous circuits. This circuit does the job using Burr-Brown's OPA111 precision monolithic dielectrically isolated (DIFET) op amp. Note the component changes needed for use with 50-Hz AC power. Source: "Integrated Circuits Data Book," Burr-Brown.

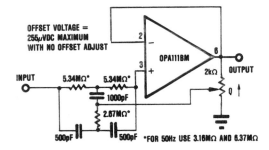

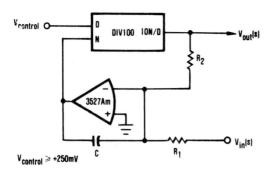

Voltage-Controlled Filter

This circuit uses a DIV100 analog divider to construct a single-pole, low-pass active filter whose cutoff frequency is linearly proportional to the circuit's control voltage. See the original publication for detailed information on the filter's transfer function and control data. Source: "Integrated Circuits Data Book," Burr-Brown.

Chapter 9
Function Generators

Function Generator
Logic Function Generator
Triangle Wave Oscillator
Triangle/Square-Wave Generator
Single-Chip Function Generator
Triangle-to-Sine Wave Converter
Waveform Generator
Function Generator

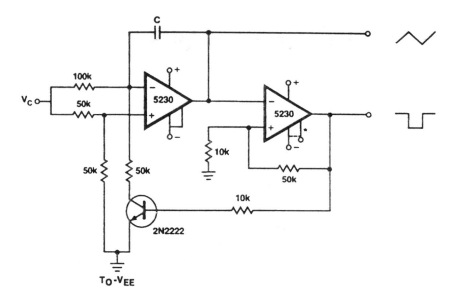

NOTE:
*See text.

Function Generator

This simple low-voltage, gated function generator uses the NE5230 and two AA batteries, so it's a perfect choice for portable applications. The outputs are square, triangular, and sine waves. The sine-wave generating circuit and the square/triangular circuits are independent. Source: "Low-Voltage Gated Function Generator: NE5230," AN1511, Linear Data Manual, Volume 2: Industrial, Signetics Corporation.

9. FUNCTION GENERATORS

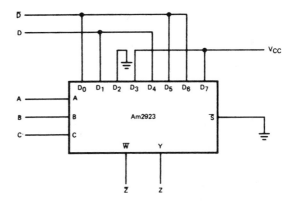

$$Z = AB\overline{C}D + \overline{A}BCD + ACD + AB + \overline{A}C\overline{D} + \overline{B}C\overline{D}$$

INPUTS				OUTPUTS	
SELECT			Output Control		
C	B	A	$\overline{S}$	Y	$\overline{W}$
X	X	X	H	.Z	Z
L	L	L	L	D_0	$\overline{D}_0$
L	L	H	L	D_1	$\overline{D}_1$
L	H	L	L	D_2	$\overline{D}_2$
L	H	H	L	D_3	$\overline{D}_3$
H	L	L	L	D_4	$\overline{D}_4$
H	L	H	L	D_5	$\overline{D}_5$
H	H	L	L	D_6	$\overline{D}_6$
H	H	H	L	D_7	$\overline{D}_7$

H = HIGH X = Don't Care
L = LOW Z = High Impedance
D_0-D_7 = The output will follow the HIGH-level or LOW-level of the selected input.
$\overline{D}_0$-$\overline{D}_7$ = The output will follow the complement of the HIGH-level or LOW-level of the selected input.

Logic Function Generator

Depending on the input, this simple circuit can generate a variety of logic functions. It uses the Am2923, an 8-bit multiplexer that switches one of eight inputs onto the inverting and noninverting outputs under the control of a 3-bit select code. As shown in the accompanying function table, this simple circuit can generate a variety of logic functions from a 3-bit select code. It uses the Am2923 8-bit multiplexer. Source: "Bipolar Microprocessor Logic and Interface Data Book," Advanced Micro Devices, Inc.

9. FUNCTION GENERATORS

Triangle Wave Oscillator

This circuit provides a variable-frequency triangular wave whose amplitude is independent of frequency. You can build it inexpensively, using just about any dual op amp, such as the EXAR XR-4558. The generator has an integrator as a ramp generator and a threshold detector with hysterious as a reset circuit. The triangular wave frequency is determined by R3, R4, and C1 and the positive and negative saturation voltages of the amplifier A1. Amplitude is determined by the ratio of R5 to the combination of R1 and R2 and the threshold detector saturation voltages. You can offset the output waveform with respect to ground if the inverting input to the threshold detector A1 is offset with respect to ground. The circuit can be made dependent of temperature and supply voltage if the detector is clamped with matched zener diodes. Source: EXAR Databook, EXAR Corporation.

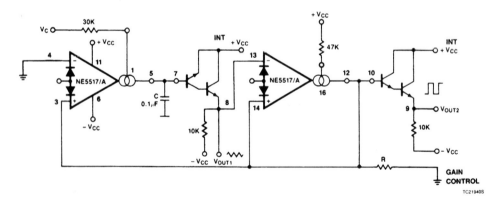

Triangle/Square-Wave Generator

This circuit uses the Signetics NE5517A dual operational transconductance amplifier to easily create a high-quality triangle and square-wave generator. Source: Linear Data Manual Volume 2: Industrial, Signetics Corporation.

9. FUNCTION GENERATORS

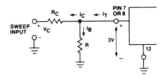

Circuit Connection for Frequency Sweep

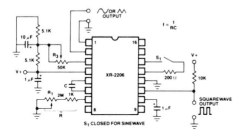

Circuit for Sine Wave Generation without External Adjustment

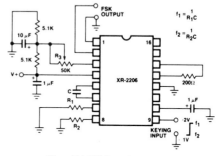

Sinusoidal FSK Generator

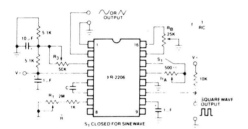

Circuit for Sine Wave Generation with Minimum Harmonic Distortion

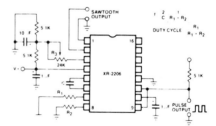

Circuit for Pulse and Ramp Generation

Single-Chip Function Generator

The EXAR XR-2206 is a monolithic function generator that can easily produce high-quality sine, square, triangle, ramp, and pulse waveforms of high stability and accuracy. The output waveforms can be both amplitude and frequency modulated by an external voltage. And you can select frequency of operation externally over a range of 0.01 Hz to more than 1 MHz. These five circuits illustrate the chip's use. See the original publication for details on selecting components. Source: EXAR Databook, EXAR Corporation.

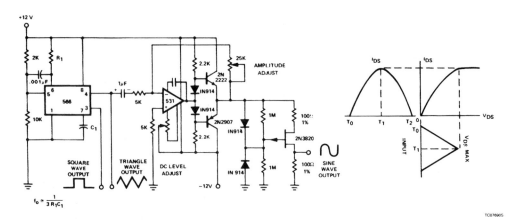

Triangle-to-Sine Wave Converter

Conversion of triangle wave shapes to sinusoids is usually done by diode–resistor shaping networks, which accurately reconstruct the sine wave segment by segment. A simpler and less costly way is shown here, using the nonlinear I_{DS}/V_{DS} transfer characteristic of a P-channel junction FET to shape the triangle waveform. The amplitude of the triangle waveform is critical and must be careful adjusted to achieve a low-distortion output. Source: "Waveform Generators with the NE566," AN186, — Linear Data Manual, Volume 1: Communications, Signetics Corporation.

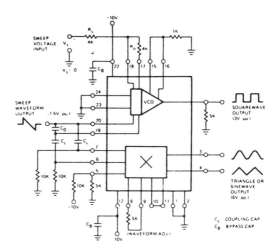

Waveform Generator

EXAR's XR-S200 multifunction PLL system chip can be interconnected to form a versatile waveform generator. This circuit generates the basic periodic square (or sawtooth) waveform. The multiplier section, connected as a linear differential amplifier, converts the differential sawtooth waveform into a triangle wave output at pins 3 and 4. You can use the waveform adjustment pot across pins 8 and 9 to round the peaks of the triangle waveform and convert it to a low-distortion sine wave. Terminals 3 and 4 can be used either differentially or single endedly to provide both in-phase and out-of-phase output waveforms. The output frequency can also be swept or frequency-modulated by applying the proper analog control input to the circuit. For linear FM modulation with relatively small frequency deviation, the modulation input can be applied across terminals 23 and 24. For large deviation sweep inputs, a negative-going sweep voltage can be applied at pin 18. This allows the frequency to be voltage-tuned over a range of approximately 10:1. The digital control inputs (pins 15 and 16) can be used for FSK applications. You can disable them by connecting them to ground through a current-limiting resistor.

Source: EXAR Databook, EXAR Corporation.

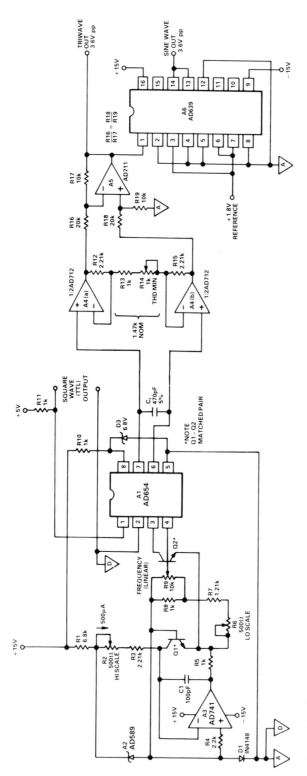

Function Generator

Since it's capable of producing the standard sine, square, and triangle waveforms, a function generator is a useful tool. The AD654 used in this circuit only produces a square wave directly from pin 1. It produces a triangle wave by differentially amplifying the capacitor waveform, and then synthesizes a sine wave from the triangle. The wide range, exponentially generated timing current for frequency control lets you use a standard linear potentiometer for a four decade control range without crowding the low end of the control span. Although R14 generally serves as an amplitude control, it's best adjusted for minimum THD on the sine wave. This circuit has a THD of 0.5% over its entire 100-kHz range. Source: Walt Jung, "Operation and Applications of the AD654 IC V-to-F Converter," Application Note E923-25-7/85, Analog Devices.

Chapter 10
Measurement Circuits

Linear Thermometer
Accelerometer
Sequential Timer
Thermocouple Amplifier
Narrow-Band Tone Detector
Negative Current Monitor
Acoustic Thermometer
Transmitting Thermometer
Magnetic Tachometer
Linear Thermometer
Voltmeter
Low-Flow-Rate Thermal Flowmeter
Presettable Timer with Linear Readout
Thermally Based Anemometer
Relative Humidity Signal Conditioners
Digital Thermometer
Sine-Wave Averaging AC Current Monitor
Level Transducer Digitizer
Thermometer Circuit
Four-Channel Temperature Sensor
Photodiode Digitizer
Field Strength Meter
Tachometer
Long-Duration Timer
Digital Thermometer

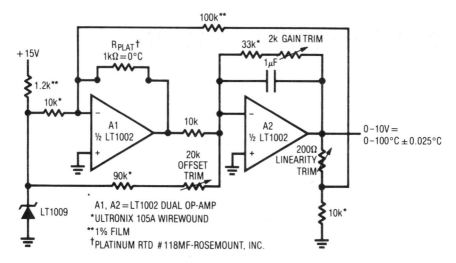

Linear Thermometer

This circuit corrects for the nonlinear temperature versus resistance characteristics of a resistance temperature detector (RTD), resulting in a thermometer that achieves an absolute accuracy of ±0.025°C over a range of 0–100°C. Calibrate this circuit by substituting a precision decade box for the sensor. Set the box to the 0°C value (100.0 ohm) and adjust the offset trim for a 0.000-V output. Next, set the decade for a 35°C output (1138.7 ohm) and adjust the gain trim for a 3.500V output reading. Finally, set the box to 1392.6 ohms (100.00°C) and trim the linearity adjustment. Repeat this sequence until all three points are fixed. Source: Jim Williams, "Applications of New Precision Op Amps," Application Note 6, Linear Technology Corporation.

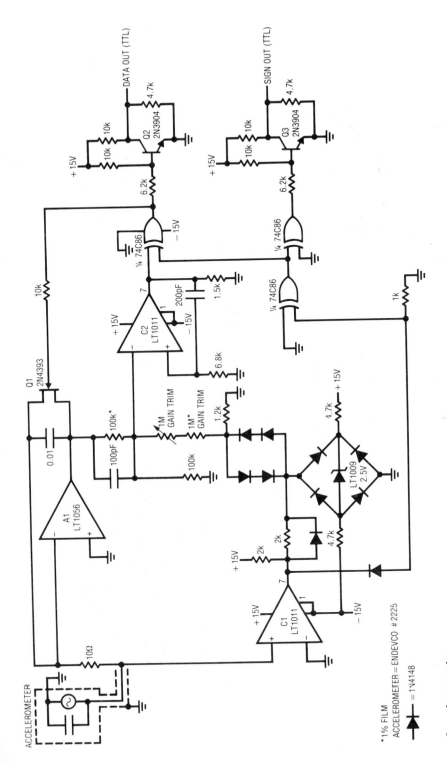

Accelerometer

A piezoelectric accelerometer relies on the property of ceramic materials to produce charge when mechanically excited. In this device a mass is coupled to the ceramic element. An acceleration acting on the mass causes charge to be dispensed from the ceramic element. This circuit accomplishes a complete, direct analog-to-digital conversion on the piezoelectric accelerometer. To trim this circuit, apply a known amplitude acceleration and adjust the 1-Mohm gain at C2. You can also electrically simulate the acceleration. (See the accelerometer manufacturer's data sheet for scale factors.) Source: Jim Williams, "Some Techniques for Direct Digitization of Transducer Outputs," Application Note 7, Linear Technology Corporation.

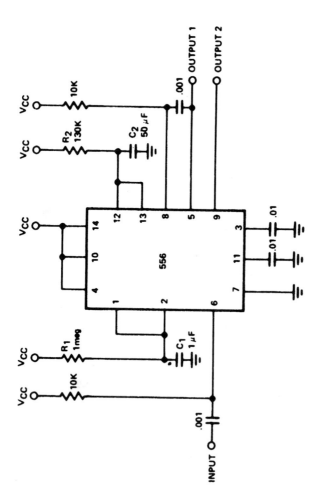

NOTE:
All resistor values are in Ω.

Sequential Timer

One feature of a dual timer such as the 556 shown here is that by utilizing both halves it's possible to obtain sequential timing. You do this by connecting the output of the first half to the input of the second half via a 0.001-μF coupling capacitor. Delay T1 is determined by the first half and delay T2 by the second half. Start the first half of the timer by momentarily connecting pin 6 to ground. When it's timed out (determined by $1.1R_1C_1$, the second half begins. Its duration is determined by $1.1R_2C_2$. Source: Linear Data Manual Volume 2: Industrial, Signetics Corporation.

10. MEASUREMENT CIRCUITS

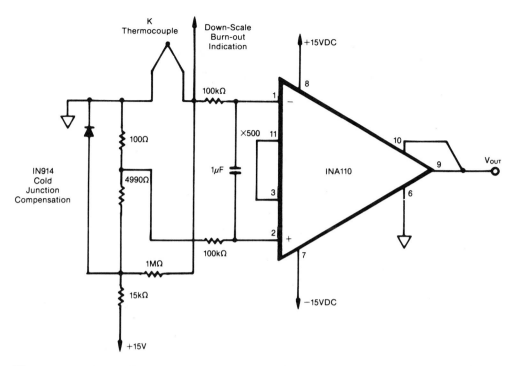

Thermocouple Amplifier

Thermocouples are used in numerous temperature-sensing applications. This amplifier uses a Burr-Brown INA110 very high-accuracy instrumentation amplifier, and includes cold-junction compensation as well as input low-pass filtering (<1Hz). Source: "Integrated Circuits Data Book," Burr-Brown.

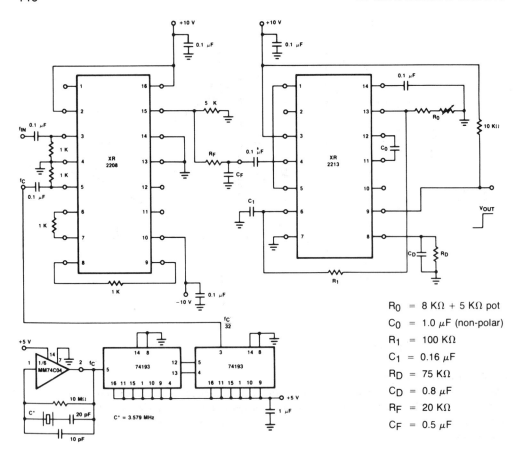

Narrow-Band Tone Detector

This circuit is capable of detecting a 1-Hz tone out of a frequency spectrum greater than 1 MHz. It can accept almost any periodic waveform including sine, square, and triangular waves. The tone detector input changes to a high state when the input is within the detection band. The circuit uses an XR-2213 PLL as the detector, along with a XR-2208 analog multiplier as a frequency mixer. Error due to temperature drift is typically 0.2%/°C. As shown here, the decoder is set for 111.7 kHz. See the original application note for detailed information on components needed for your desired frequency range. Source: "Precision Narrow-Band Tone Detector," AN-21, EXAR Databook, EXAR Corporation.

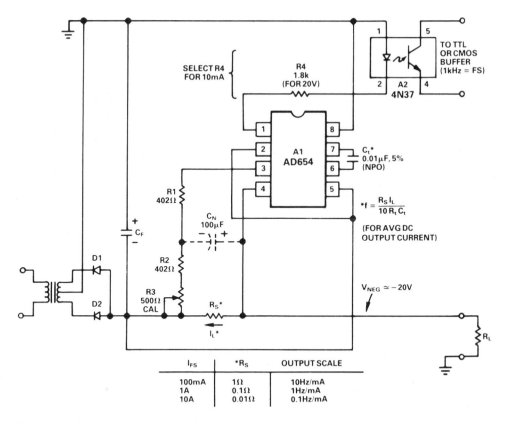

Negative Current Monitor

This circuit illustrates a technique for monitoring negative supply current, in this case a −20-V bus. In this hookup, the negative voltage input mode is used, with the V/F measuring the drop across R_s, a series sampling resistor. You should select the value of R_s from the table for the full-scale current that you require. Source: Walt Jung, "Operation and Applications of the AD654 IC V-to-F Converter," Application Note E923-25-7/85, Analog Devices.

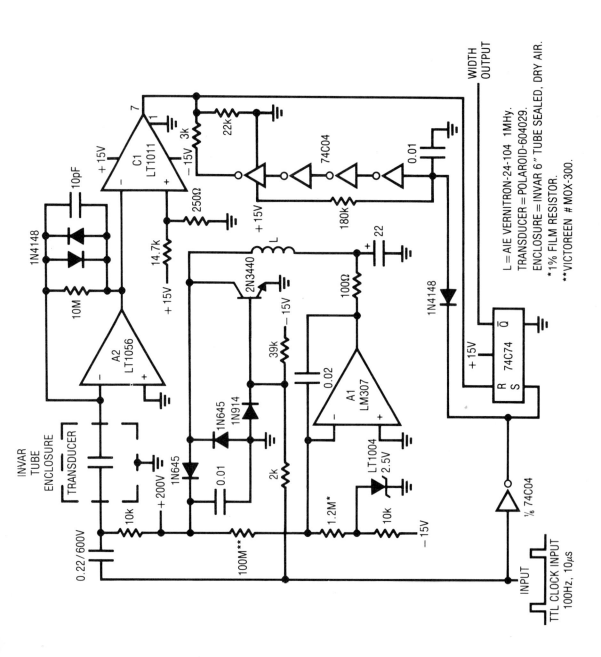

Acoustic Thermometer

This circuit uses an unusual transducer to measure temperature, using the relationship between the sound and the temperature in dry air. This acoustic thermometry is used where extremes in operating temperatures are encountered, such as cryogenics or nuclear reactors. The transducer in this circuit is composed of a Polaroid ultrasonic element mounted at one end of a sealed, 6-inch long invar tube filled with dry air. Source: Jim Williams, "Some Techniques for Direct Digitization of Transducer Outputs," Application Note 7, Linear Technology Corporation.

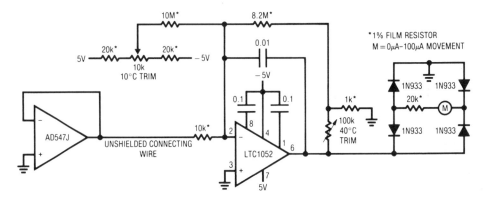

Transmitting Thermometer

This thermometer is unique because it allows the temperature transducer's output to be directly transmitted over an unshielded wire. In this case, the circuit is arranged for a +10– +40°C output range. To calibrate, subject the op amp to a +10°C environment and adjust the 10°C trim for an appropriate meter indication. Next, place the op amp sensor in a +40°C environment and trim the 40°C adjustment for the proper reading. Repeat this procedure until both points are fixed. This circuit will typically provide accuracy with ±■°C, even in high noise environments. Source: Jim Williams, "Application Considerations and Circuits for a New Chopper-Stabilized Op Amp," Application Note 9, Linear Technology Corporation.

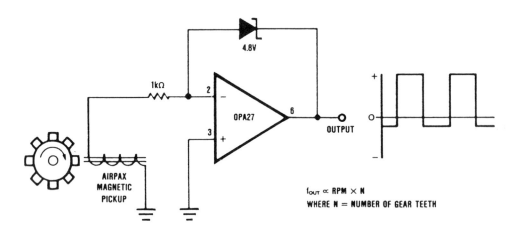

Magnetic Tachometer

This simple circuit uses Burr-Brown's OPA37A ultra-low noise op amp along with a standard Airpax magnetic pickup to output the speed of a gear. Source: "Integrated Circuits Data Book," Burr-Brown.

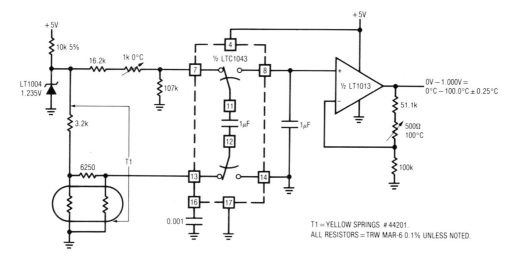

Linear Thermometer

This circuit uses a thermistor network to eliminate the requirement for a linearity trim. However, it's at the expense of accuracy and range of operation. Source: Jim Williams, "Applications for a Switched-Capacitor Instrumentation Building Block," Application Note 3, Linear Technology Corporation.

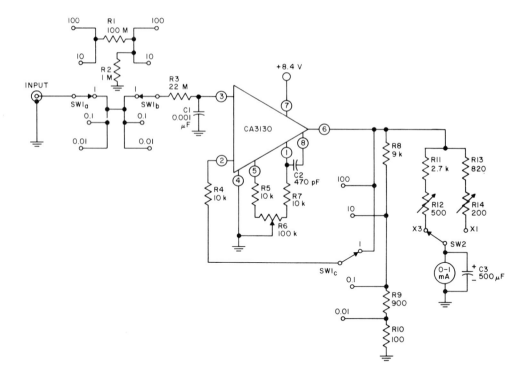

Voltmeter

This high-input resistance voltmeter exploits a number of the inherent characteristics of the RCA CA3130 BiMOS op amp. Fundamentally, it's a single-supply voltage-follower circuit with sufficient gain to drive a 1-milliampere meter in response to picoampere-input currents. With the resistor–divider network (R1,R2), it can measure DC voltages over a range of 10 millivolts to 300 volts. Calibration resistors R12 and R14 individually establish full-scale meter deflection when 1 milliampere is forced to flow through resistors R11, R12, and R14. Op amp and meter nulling is done by first setting the range selector switch SW1 to the lowest voltage position and shorting the input terminal to ground. Then adjust nulling potentiometer R6 until you see the first indication of meter movement. Source: H.A. Wittlinger, "Understanding and using the CA3130, CA3130A, and CA3130B BiMOS Operational Amplifiers," Linear Integrated Circuits Application Note ICAN-6386, RCA Solid State Division.

10. MEASUREMENT CIRCUITS

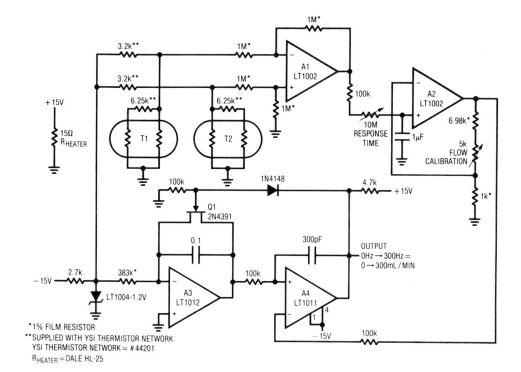

*1% FILM RESISTOR
**SUPPLIED WITH YSI THERMISTOR NETWORK
YSI THERMISTOR NETWORK = #44201
R_{HEATER} = DALE HL-25

Low-Flow-Rate Thermal Flowmeter

Measuring flow rates in slow-moving liquids is difficult. Paddle wheel and hinged-vane transducers are inaccurate in these cases. And they're impractical with small-diameter tubing. This circuit is a thermally based flowmeter that features high accuracy at rates as low as 1 ml per minute. It has a frequency output that's a linear function of flow rate. It uses two sensors, the first of which assumes the fluid's temperature before it's heated by the resistor. The second sensor picks up the fluid's heat rise. Source: Jim Williams, "Thermal Techniques in Measurement and Control Circuitry," Application Note 5, Linear Technology Corporation.

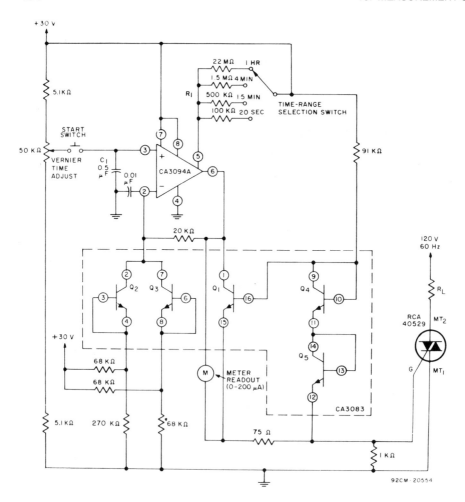

Presettable Timer with Linear Readout
In many timer applications, you need a meter readout of the actual elapsed time. This circuit uses the RCA CA3094 monolithic programmable power switch/amplifier and CA3083 transistor array to control the meter and a load-switching triac. Current flow in the meter is essentially linear with respect to the timing period. Source: L. R. Campbell and H. A. Wittlinger, "Some Applications of a Programmable Power Switch/Amplifier." Application Note ICAN-6048, RCA Solid State.

10. MEASUREMENT CIRCUITS

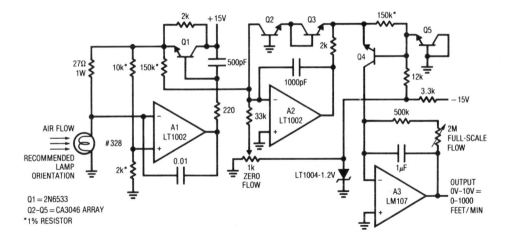

Thermally Based Anemometer

This air (or gas) flowmeter works by measuring the energy required to maintain a heated resistance wire at a constant temperature. It's accurate within 3% over its entire 0–1000 foot per minute range. A type 328 lamp is modified for this circuit by removing its glass envelope. To use this circuit, place the lamp in the air flow so that its filament is at a 90° angle to the flow. Next, either shut off the air flow or shield the lamp from it and adjust the zero flow potentiometer for a circuit output of 0 V. Then expose the lamp to an air flow of 1000 feet per minute and trim the full flow potentiometer for 10-V output. Repeat these adjustments until both points are fixed. Source: Jim Williams, "Thermal Techniques in Measurement and Control Circuitry," Application Note 5, Linear Technology Corporation.

10. MEASUREMENT CIRCUITS

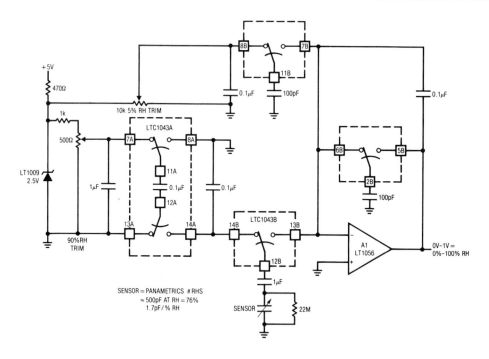

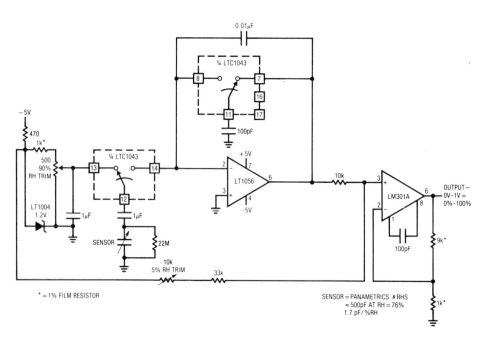

10. MEASUREMENT CIRCUITS

◄ Relative Humidity Signal Conditioners

Relative humidity is a difficult physical parameter to measure, and most of the transducers available for it require fairly complex signal conditioning circuitry. This first circuit makes things simpler by combining two LTC1043s along with a capacitively based humidity transducer in a simple charge-pump circuit. To calibrate this circuit, place the sensor in a known 5% relative humidity environment and adjust the "5% RH trim" for 0.05-V output. Next, place the sensor in a known 90% relative humidity environment and set the "90% RH trim" for 900-mV output. Repeat this procedure until both points are fixed. Once calibrated, this circuit is accurate within 2% in the 5–90% relative humidity range. The second circuit is an alternate that requires two op amps but only one LTC1043 package. This circuit retains insensitivity to clock frequency while permitting a DC offset trim. Source: Jim Williams, "Applications for a Switched-Capacitor Instrumentation Building Block," Application Note 3, Linear Technology Corporation.

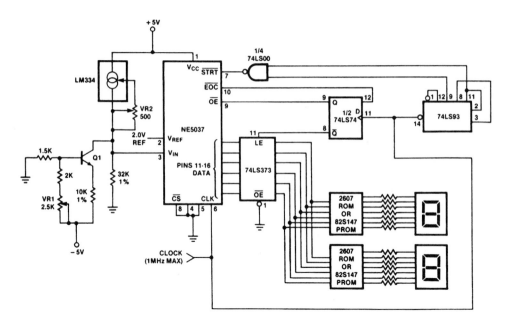

Digital Thermometer

Here's a typical and low-cost application for a circuit like the Signetics NE5037 6-bit A/D converter. The LM334 is used here as the temperature sensor, and the NE5037 provides the A/D conversion to the display circuitry. Source: Linear Data Manual Volume 2: Industrial, Signetics Corporation.

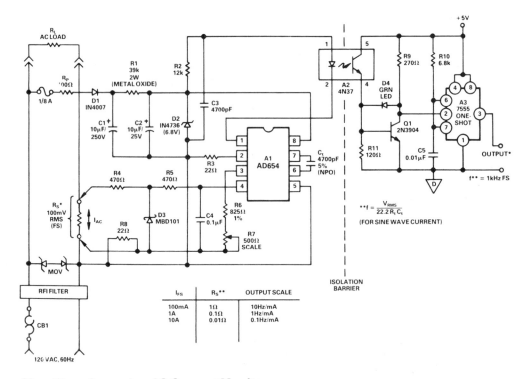

Sine-Wave Averaging AC Current Monitor

This circuit is self powered and measures the average AC line current in the load R_L, delivering an isolated output frequency. Although as shown, this circuit measures current, you can rewire it to measure line voltage by adding a divider to produce 100-mV rms AC at R4. This circuit can be either line powered by a peak detector circuit and series resistor (D1, C1, R10), or you can power it from a small rectifier transformer and a simple half-wave supply. Although it functions adequately when self powered, it's not very efficient from a power consumption standpoint. If powered from the AC line, it's extremely important that you safety precautions, including the use of overvoltage clamps, breakers, sufficient derating, etc. Source: Walt Jung, "Operation and Applications of the AD654 IC V-to-F Converter," Application Note E923-25-7/85, Analog Devices.

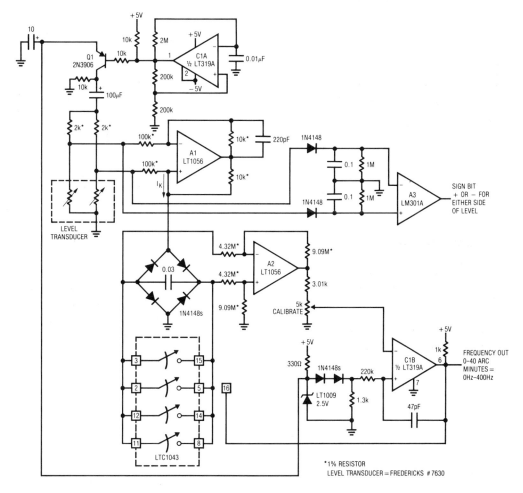

Level Transducer Digitizer

Level transducers measure the angle away from true level, and are employed in road construction, machine tools, inertial navigation systems, and other applications that require a gravity reference. This circuit uses the elegantly simple level transducer of a small tube nearly filled with a partially conductive liquid. If the tube is level with respect to gravity, the bubble resides in the tube's center and the electrode resistances to common are identical. As the tube shifts away from level, the resistances increase and decrease proportionally. This circuit directly produces a calibrated frequency output corresponding to level. A sign bit, also supplied at the output, gives polarity information. To calibrate this circuit, place the level transducer at a known 40 arc-minute angle and adjust the 5-kohm trimmer at C1B for a 400-Hz output. Circuit accuracy is limited by the transducer to about 2.5%. Source: Jim Williams, "Some Techniques for Direct Digitization of Transducer Outputs," Application Note 7, Linear Technology Corporation.

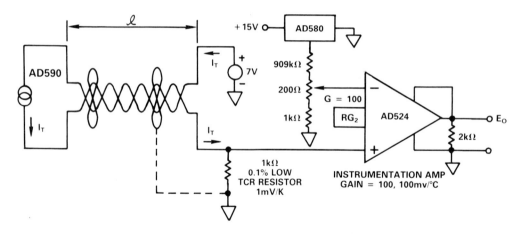

Thermometer Circuit

The AD590 is a two-terminal IC temperature transducer that produces an output proportional to absolute temperature. In this typical application, it's used as a remote temperature-to-current transducer that measures temperatures from −55°C to 100°C. Since the AD590 measures absolute temperature, its output must be offset for measuring degrees centigrade. The resistors and AD580 perform this task. See the original application note for useful information on eliminating noise in remote pickup cables. Source: Paul Klonowski, "Use of the AD590 Temperature Transducer in a Remote Sensing Application," Application Note E920-15-6/85, Analog Devices.

10. MEASUREMENT CIRCUITS

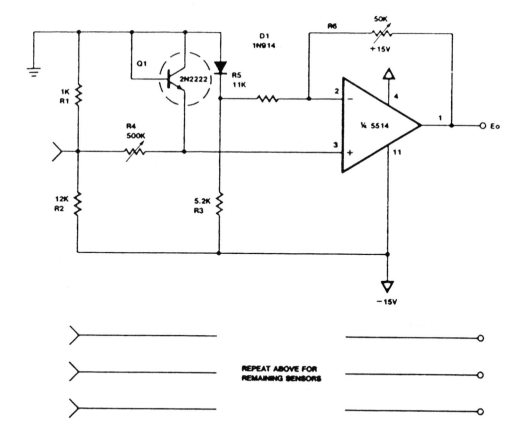

Four-Channel Temperature Sensor

By using an NPN transistor as a temperature sensing element, the Signetics NE/SE5514 quad op amp forms the basis for a multistation temperature sensor. The principle used is fundamental to the current voltage relationship of a forward-biased junction. Diode D1 serves to substantially reduce error due to power-supply variation by giving a fixed voltage reference. To calibrate the sensor, adjust R4 for 0-V output from the NE5514 at 0°C. Then adjust the R6 tracking resistor for a scale factor of 100-V/°C output. Only the sensing transistors need to be placed in a temperature-controlled environment. The second circuit shows how to add an A/D converter to give you a digital thermometer. Source: "Applications of the NE5514," AN1441, Linear Data Manual Volume 2: Industrial, Signetics Corporation.

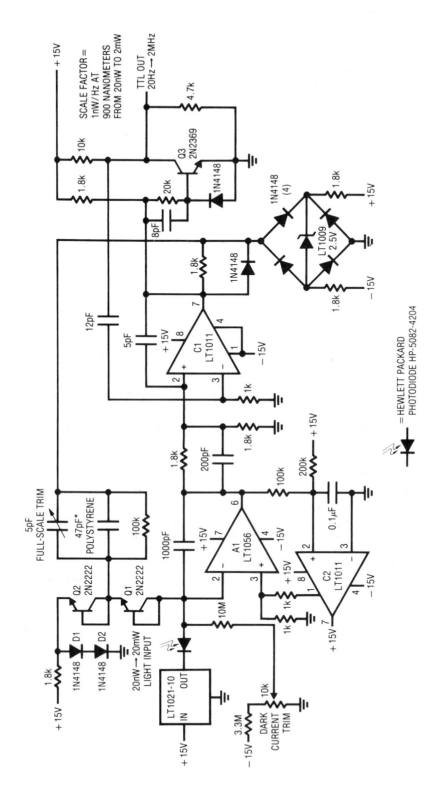

Photodiode Digitizer

Because of their extremely wide dynamic range, photodiodes present a difficult challenge for signal conditioning circuitry. Since a high-quality photodiode furnishes a linear current output over a 100-dB range, it requires a 17-bit analog-to-digital converter as well as a current-to-voltage input amplifier. This circuit directly converts a photodiode's current output into an output frequency with a 100-dB dynamic range. Optical input power of 20 nW to 2 mW produces a linear, calibrated 2-Hz to 2-MHz output. Output response to input light steps is fast and the cost is low. To trim this circuit, place the photodiode in a completely dark environment. Trim the dark current adjustment so the circuit oscillates at the lowest possible frequency. Next, apply or electrically simulate a 2-mW optical input. Trim the 5-pF adjustment for an output frequency of 2 MHz. If the adjustment is outside the range of the trimmer, alter the 47-pF capacitor's value accordingly. Once calibrated, this circuit will maintain a 1% accuracy over the photodiode's entire 100-dB range. Source: Jim Williams, "Some Techniques for Direct Digitization of Transducer Outputs," Application Note 7, Linear Technology Corporation.

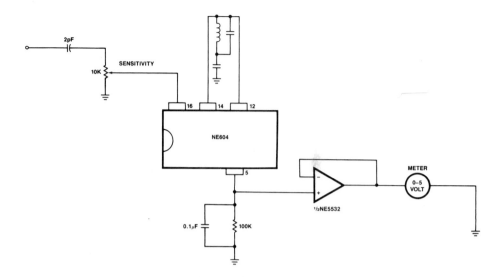

Field Strength Meter

Although Signetics' NE/SA604 FM IF chip was designed for cellular radio applications, its features make it ideal for numerous other applications. The chip has an IF amplifier, quadrature detector, received signal strength indicator, and a mute circuit. This circuit makes use of its signal strength feature, for a simple and accurate FM field strength meter. Source: "Designing with the NE/SA604." AN-199, Linear Data Manual, Volume 1: Communications, Signetics Corporation.

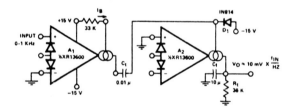

Tachometer

This simple frequency-to-voltage converter uses EXAR's XR-13600 dual operational transconductance amplifier. Source: EXAR Databook, EXAR Corporation.

10. MEASUREMENT CIRCUITS

Long-Duration Timer

This timer circuit can provide a time delay of up to 20 minutes. The circuit is a standard relaxation oscillator with an FET current source in which resistor R1 is used to provide reverse bias on the gate-to-source of the JFET. This turns the JFET off and increases the charging time of C1. C1 should be a low-leakage capacitor such as a mylar type. Source: R. J. Haver and B. C. Shiner, "Theory, Characteristics and Applications of the Programmable Unijunction Transistor," Motorola Semiconductor Products Application Note AN-527, Copyright Motorola, Inc. Used by permission.

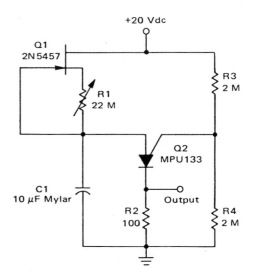

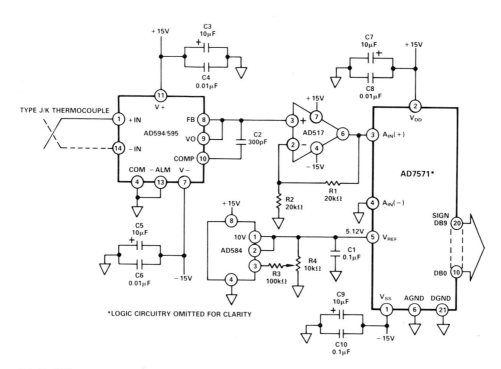

Digital Thermometer

This circuit uses a combination of the AD7571 analog-to-digital converter along with the AD594/595 thermocouple amplifier, and is easy to interface to a microprocessor. It's precalibrated to match both type-J and type-K thermocouples. A scaling amplifier is used at the AD594/595 output to ensure that the full dynamic range of the A/D converter is used. With a scaling amplifier gain of 2, this circuit's resolution is 0.5°C, and the temperature range is ±512°C. Source: John Reidy, "Temperature Measurement System to 10-Bit Resolution Using the AD7571 and the AD594/595." Application Note E922-15-7/85, Analog Devices.

Chapter 11
Microprocessor Circuits

8-Bit Serial-to-Parallel Converter
Voltage-Sag Detector
Power-Loss Detection Circuit
CRT Driver
Clock Regenerator
Bidirectional Bus Interface
Cheapernet/Ethernet Interface

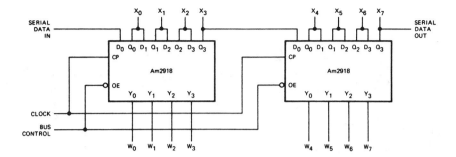

8-bit Serial-to-Parallel Converter

The Am2917A is a high-performance, low-power Schottky bus transceiver that consists of four D-type, edge-triggered flip flops. Their outputs are connected to four 3-state bus drivers, with each driver internally connected to the input of a receiver. This circuit shows a typical application. It's an 8-bit serial-to-parallel converter that has 3-state output (W) and direct access to the register word (X). Source: "Bipolar Microprocessor Logic and Interface Data Book," Advanced Micro Devices, Inc.

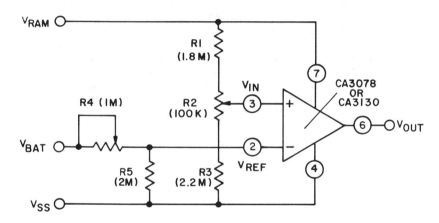

Voltage-Sag Detector

This circuit is used with CMOS RAMs to detect a voltage sag. Some precautions must be taken to assure that V_{DD} doesn't fall below an acceptable system operating voltage. The circuit constantly monitors the V_{RAM} voltage and compares it to the acceptable minimum system voltage. With this detector, you can place a RAM array on battery-backup operation. Source: Carmine Salerno, "Application of the CA1524 Series Pulse-Width Modulator ICs," Linear Integrated Circuits, Application Note ICAN-6915, RCA Solid State Division.

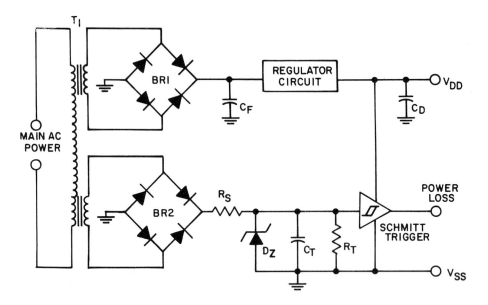

Power-Loss Detection Circuit

In microprocessor systems, some advanced warning in the event of an unscheduled AC power loss is imperative. Enough time must be provided to allow shutdown of normal system operations at a convenient breakpoint in a program. Since the output of the V_{DD} supply bridge rectifier (BR1) is followed by a large filter capacitor (C_F) and since the V_{DD} line itself usually has some distributed capacitance, the V_{DD} supply will remain stable for a long enough time after the power-loss signal is issued to effect an orderly system shutdown. Source: R.M. Vaccarella, "Designing Minimum/Nonvolatile Memory Systems with CMOS Static RAMs." Memory Microprocessor Products Application Note ICAN-6943, RCA Solid State Division.

11. MICROPROCESSOR CIRCUITS

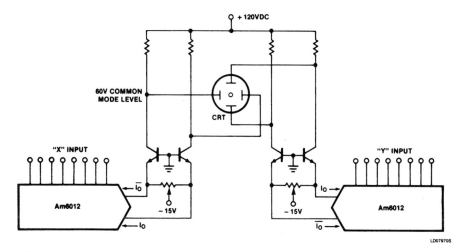

NOTES:
1. Full differential drive lowers power supply voltage.
2. Eliminates inverting amplifiers and transformers.
3. Independent beam centering controls.

CRT Driver

The Signetics AM6012 12-bit multiplying digital-to-analog converter uses a 3-bit segment generator for the MSBs in conjunction with a 9-bit R-2R diffused resistor ladder to provide 12-bit resolution. While this device requires a reference input of 1 mA for a 4 mA full-scale current, its operation is nearly independent of power supply shifts. This circuit shows a typical computer application with the IC directly driving the X–Y input of a CRT. Source: Linear Data Manual Volume 2: Industrial, Signetics Corporation.

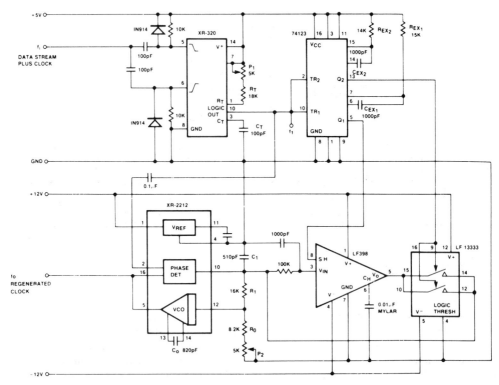

P1 = ADJUST SO POSITIVE PORTION OF f1 IS EQUAL TO ½ OF THE CLOCK PERIOD
P2 = ADJUST FOR 90 PHASE SHIFT BETWEEN f1 and fo WITH f1 = fCLK

Clock Regenerator

Recovering encoded serial data from floppy disk systems poses a major design problem as the synchronized clock used to encode data is embedded within the data stream. The clock can't be readily extracted using common phase-locked loop techniques since the actual clock may appear for only short periods of time in a common encoding format such as NRZI. This clock is necessary to decode the serial data and retrieve the original data. This PLL circuit can be used to recover the clock from a serial data stream using NRZI protocol with excellent stability. It uses the XR-2122 precision phase-locked loop in conjunction with the XR-320 monolithic timer to form the heart of the system. A 74123 dual one-shot and a 398/13333 are used for timing and sample and hold purposes. Source: "Clock Recovery System," AN-19, EXAR Databook, EXAR Corporation.

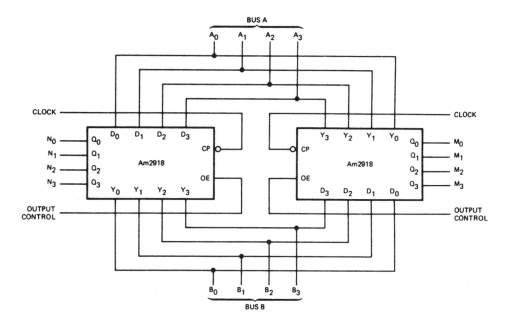

Bidirectional Bus Interface

The Am2917A bus transceiver can be connected as shown to act as a bidirectional interface between two buses. The device on the left both stores data from the A-bus and drives it. The device on the right stores data from the B-bus and drives the A-bus. You use the output control to place either or both drivers into the high-impedance state. The contents of each register are available for continuous usage at the N and M ports of the device. Source: "Bipolar Microprocessor Logic and Interface Data Book," Advanced Micro Devices, Inc.

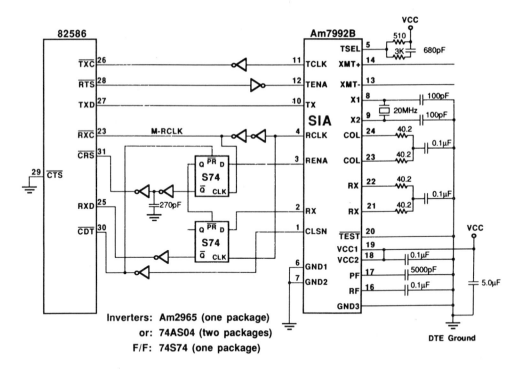

Cheapernet/Ethernet Interface

The Cheapernet local-area network is essentially the industry-standard Ethernet with cheaper cables, fewer nodes, and shorter distances between nodes. But it still provides the same 10 megabit per second performance. The Am7990 family is compatible with IEEE 802.3, the industry standard for Ethernet/Cheapernet. This circuit shows how to interface AMD's Cheapernet controller and Intel's Ethernet controller. See the original reference guide for detailed information on use. Source: "Am7990 Ethernet/Cheapernet Family Reference Guide," Advanced Micro Devices, Inc.

Chapter 12
Miscellaneous Circuits

Intercom
Buffered Output Line Driver
Low-Voltage Lamp Flasher
Digital Clock with Alarm
Astable Single-Supply Multivibrator
Frequency Synthesizer
Fed Forward, Wideband DC-Stabilized Buffer
Three Clock Sources
Freezer Alarm
Fiber Optic Receiver
Micropower Sample-Hold
Frequency Output Analog Divider
Protected High Current Lamp Driver
Tone Transceiver
Variable Shift Register
Fast, Precision Sample-Hold Circuit
Precision PLL
Two Peak Circuits

Intercom

This circuit integrates a simple low-power audio amplifier into a low-cost intercom by adding only speakers, a switch, and an input transformer. In this design, the listen/talk position switch controls two or more remote positions. If you'll only be using the intercom intermittently, a standard 9-volt alkaline cell makes a suitable power supply. Source: W. M. Austin and H. M. Kleinman, "Application of the RCA CA3020 and CA3020A Integrated-Circuit Multi-Purpose Wide-Band Power Amplifiers," Application Note ICAN-5766, RCA Solid State.

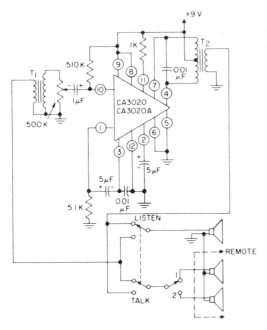

T_1: Primary 4 ohms,
Secondary 25,000 ohms;
Stancor A4744 or equiv.

T_2: Better Coil and Transformer DF1084, Thordarson TR-192, or equiv.

Speakers: 4 ohms

Low-Voltage Lamp Flasher ▶

One advantage of a programmable unijunction transistor (PUT) over a conventional unijunction transistor is that the PUT operates very well for low supply voltages. This is due to the low forward voltage drop of the PUT, 1.5-V maximum compared to the emitter saturation voltage of a unijunction transistor of a typical 3-V maximum. This circuit shows a low-voltage lamp flasher composed of a relaxation oscillator formed by Q1 and SCR flip–flop formed by Q2 and Q3. Note that C4 is a nonpolarized capacitor. For the component values shown, the lamp is on for about 1/2 second and off for the same amount of time. Source: R. J. Haver and B. C. Shiner, "Theory, Characteristics and Applications of the Programmable Unijunction Transistor," Motorola Semiconductor Products, Application Note AN-527, Copyright of Motorola, Incorporated. Used by Permission.

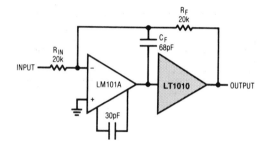

Buffered Output Line Driver

This circuit uses the Linear Technology LT1010 power buffer within the feedback loop of an operational amplifier. At lower frequencies, the buffer is within the feedback loop and its offset voltage and gain error are negligible. At higher frequencies, feedback is through C_F so that phase shift from load capacitance acting against the buffer's output resistance doesn't cause loop instability. The speed of this circuit is limited by the op amp. Source: Jim Williams, "Applications for a New Power Buffer," Application Note 4, Linear Technology Corporation.

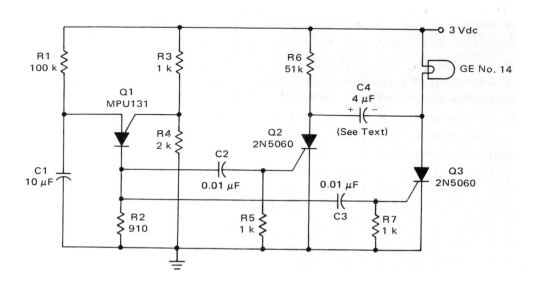

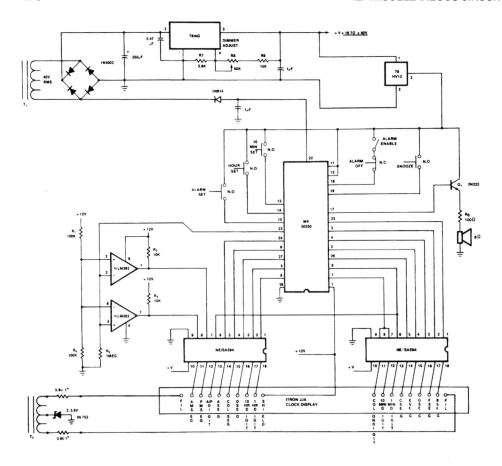

Digital Clock with Alarm

This full-featured clock is constructed with a few chips and some discrete components. Although the MK50250 performs the timekeeping and alarm functions, take note of the dual Signetics NE/SA594 display drivers, which are the heart of the display system. The chips drive any vacuum fluorescent displays, and have inputs designed to be compatible with TTL, DTL, NMOS, PMOS, or CMOS output circuitry. Since there's an active pull-down circuit on each output, the chips minimize the common problem of display ghosting. Source: Linear Data Manual, Volume 2: Industrial, Signetics Corporation.

12. MISCELLANEOUS CIRCUITS

Astable Single-Supply Multivibrator

This circuit uses the RCA CA3094 monolithic programmable power switch/amplifier to flash an incandescent lamp. With the component values shown, the circuit produces one flash per second with a 25% on time while delivering output current in excess of 100 milliamperes. During the 75% off time, it idles with micropower consumption. The flashing rate can be maintained within ±2% of the nominal value over a battery voltage range of 6 to 15 volts, and a temperature range from 0 to 70°C. Source: L. R. Campbell and H. A. Wittlinger, "Some Applications of a Programmable Power Switch/Amplifier." Application Note ICAN-6048, RCA Solid State.

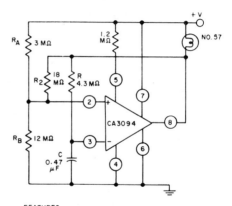

FEATURES
- 1 FLASH/SEC. $f_{OSC} = \dfrac{1}{2RC \ln[(2R_1/R_2 + 1)]}$ WHERE $R_1 = \dfrac{R_A R_B}{R_A + R_B}$
- 25% DUTY CYCLE
- FREQUENCY INDEPENDENT OF V+ FROM 6-15 V DC

Frequency Synthesizer

This frequency synthesizer uses the VCO section of EXAR's XR-S200 multifunction PLL system chip. Actual frequency synthesis is performed by a PLL closed with a programmable counter or digital divide-by-N circuit inserted into the feedback loop. The VCO signal is divided by N, so that when the circuit locks to an input signal at frequency f_s, the oscillator output is Nf_s. By changing N, you can synthesize a large number of discrete frequencies from a given reference frequency. Source: EXAR Databook, EXAR Corporation.

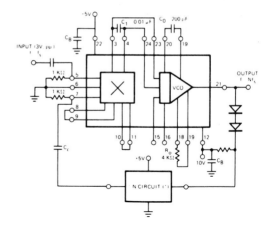

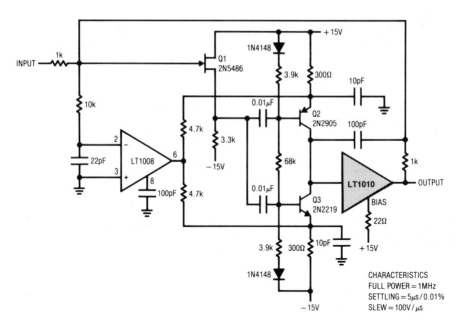

Fed Forward, Wideband DC Stabilized Buffer

This circuit eliminates the speed limitation of an operational amplifier used in a buffer. The LT1010 power buffer is combined with a wideband gain stage (Q1-Q3) to form a fast inverting configuration. The high speed is the result of the DC stabilization path occurring in parallel with the buffer. Source: Jim Williams, "Applications for a New Power Buffer," Application Note 4, Linear Technology Corporation.

12. MISCELLANEOUS CIRCUITS

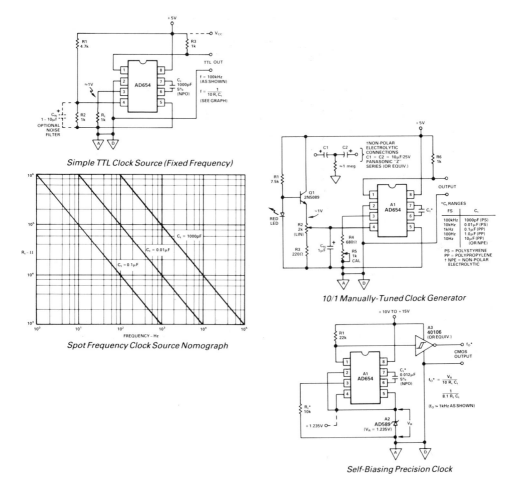

Simple TTL Clock Source (Fixed Frequency)

Spot Frequency Clock Source Nomograph

10/1 Manually-Tuned Clock Generator

Self-Biasing Precision Clock

Three Clock Sources

The AD654 V/F converter is useful as a stand-alone clock source because of its ease of logic output adaptation, low power consumption, and ease of tuning (when needed). These three circuits illustrate different approaches. — Circuit a is one of the simplest clock sources possible. Note that even if the circuit is powered from +5V, the output can be pulled to higher voltages such as 10 or 15 V. It's most useful when you need a single spot frequency with a single R-C combination, as shown in the nomograph. — Circuit b is adapted for manual tuning and switched capacitors. It offers smooth tuning over a 10/1 range. — Circuit c improves the previous two circuits' lack of high precision and lack of immunity to power supply changes. If exact frequency operation is needed, you'll need to calibrate the circuit. Source: Walt Jung, "Operation and Applications of the AD654 IC V-to-F Converter," Application Note E923-25-7/85, Analog Devices.

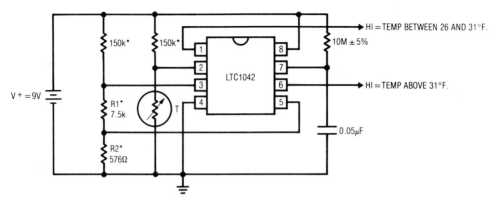

Freezer Alarm

Circuits such as this one are used in industrial and home freezers, as well as in refrigerated trucks and rail cars. It uses an LTC1042 sampled operation window comparator. The R-C combination sets a sample rate of 1 Hz, and the bridge values program the internal window comparator for the outputs shown. For normal freezer operation, pin 1 is high and pin 6 is low. Over temperature reverses this state and can trigger an alarm. The circuit consumes about 80 µA of current.
Source: Jim Williams, "Micropower Circuits for Signal Conditioning," Application Note 23, Linear Technology Corporation.

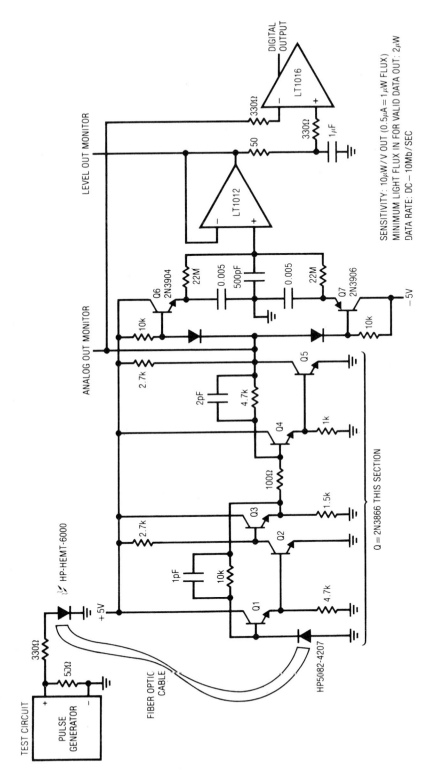

Fiber Optic Receiver

Receiving high-speed fiber optic data isn't easy. The speed of data and uncertain intensity of the light level can cause erroneous results unless the receiver is carefully designed. This fiber optic receiver will accurately condition a wide range of light inputs at data rates up to 10 MHz. Its digital output features an adaptive threshold trigger that accommodates varying signal intensities due to component aging and other causes. An analog output is also available to monitor the detector output. Source: Jim Williams, "High Speed Comparator Techniques," Application Note 13, Linear Technology Corporation.

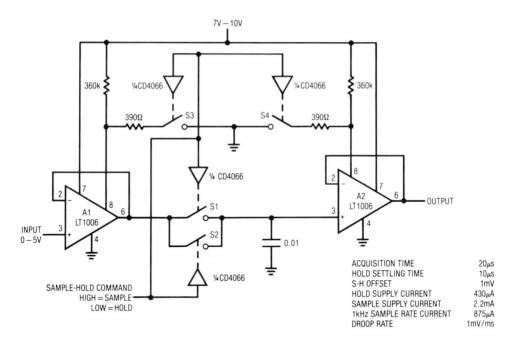

Micropower Sample-Hold

This low-power sample-hold circuit has an acquisition time of 20 μs. It takes advantage of the programming pin on the LT1006 op amp to maximize speed and performance. In normal operation, the sample time is short to hold, and the current consumption is low. Source: Jim Williams, "Micropower Circuits for Signal Conditioning," Application Note 23, Linear Technology Corporation.

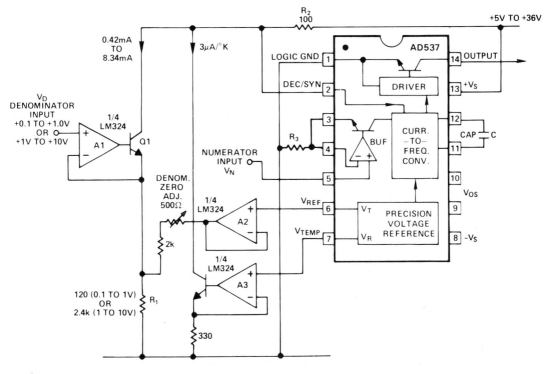

Frequency Output Analog Divider

This circuit generates a frequency proportional to the ratio of two voltages. To adjust the denominator offset, connect the V_N and V_D inputs together and trim to maintain frequency independent of input voltage. Source: Doug Grant, "Applications of the AD537 IC Voltage-to-Frequency Converter," Application Note E478-10-8/78, Analog Devices.

Protected High Current Lamp Driver

In this application, the Linear Technology LT1083, which has the ability to handle loads up to 7.5 A, is used to good advantage to drive a 12-V, 5-A lamp. Source: "Linear Databook Supplement," Linear Technology Corporation.

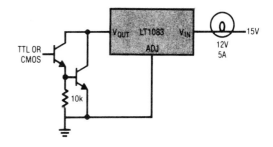

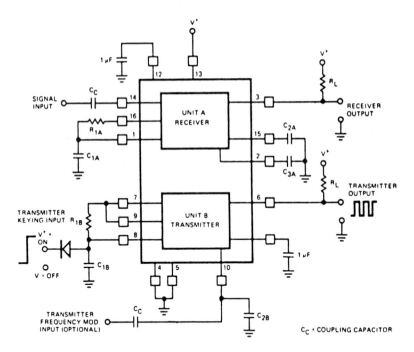

Tone Transceiver

The EXAR XR-2567 dual monolithic tone decoder has an oscillator output that you can use as a full-duplex tone transceiver by using one section of the unit as a tone detector and the remaining section as a tone generator. Since both sections operate independently, you can use this circuit to transmit and receive simultaneously. As shown here, Unit A is the receiver and Unit B is the transmitter. You can key the transmitter section on and off by applying a pulse to pin 8 through the disconnect diode D1. The oscillator section of Unit B is keyed off when the keying logic level at pin 8 is at a low state. The output of the transmitter section (Unit B) can also be modulated over a ±6% deviation range by applying a modulation signal to pin 10. Source: EXAR Databook, EXAR Corporation.

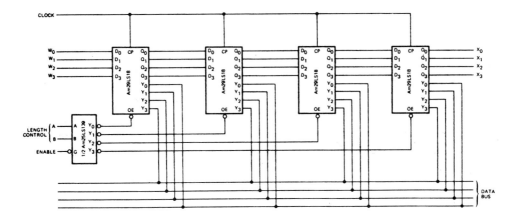

Variable Shift Register

The AM29LS10 consists of four D-type flip–flops with a buffered common clock. Essentially, it's a 4-bit, high-speed register that's particularly useful in real-time signal-processing applications. In this example, four Am29LS10s are used as a variable length (1,2,3, or 4 word) shift register. Source: "Bipolar Microprocessor Logic and Interface Data Book," Advanced Micro Devices, Inc.

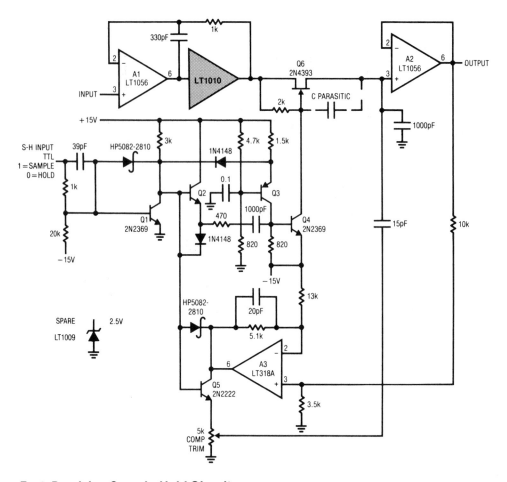

Fast, Precision Sample-Hold Circuit

Typical sample-hold circuits require high capacitive load driving capability to achieve fast acquisition times. Also required are high charge currents and dynamic stability, which the LT1010 power buffer can provide. This fast and precise sample-hold circuit has an acquisition time of 2 µs, a hold setting time of <100 nsec to 1 mV, and an aperture time of 16 nsec. Source: Jim Williams, "Applications for a New Power Buffer," Application Note 4, Linear Technology Corporation.

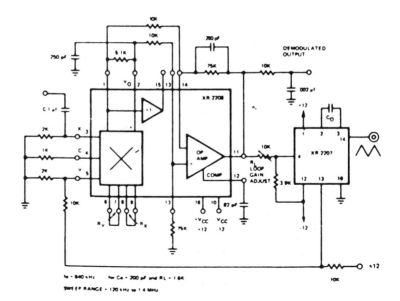

Precision PLL

This precision PLL is constructed from an XR-2207 VCO and an XR-2208 multiplier. Because of the temperature stability and wide sweep range of the XR-2207, this PLL exhibits especially good stability of center frequency and wide lock range. In this application the XR-2208 serves as a phase comparator and level shifter. Resistor R_L adjusts the loop gain of the PLL, thus varying the lock range. You can vary the tracking range from about 1.4:1 up to 12:1. Source: EXAR Databook, EXAR Corporation.

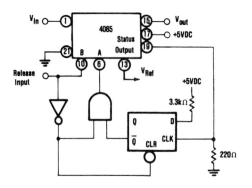

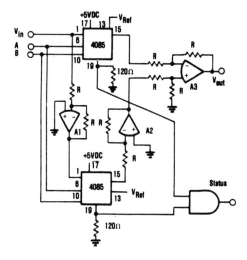

Two Peak Circuits

Both these circuits use a Burr-Brown 4085 peak detector. The first, a peak catcher, detects and holds the first peak it encounters. To reset the circuit for catching another peak, a 10-μS or longer positive logic pulse should occur at the release input. The second circuit will display the peak-to-peak voltage of an input waveform. The status output indicates that both positive and negative peaks have been detected and that the output is valid. To insure good common-mode rejection, you should match the resistors around A3. Source: "Integrated Circuits Data Book," Burr-Brown.

Chapter 13
Modem Circuits

Bell 212A Modem
300-bps Full-Duplex Modem
1200-bps Modem
Full-Duplex 300/1200 bps Modem System
2400-bps CCITT Modem
Auto Dialer Modem
2400-bps Stand-Alone Intelligent Modem
Notch Filters
Full-Duplex FSK Modem
High-Speed FSK Modem
Power Line Modem

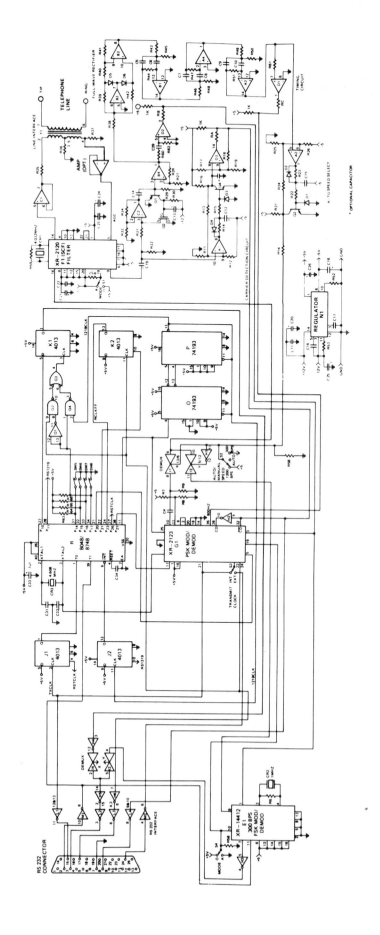

A	XR-4741	Quad Op Amp	
B	XR-4741	Quad Op Amp	
C	XR-1458	Dual Op Amp	
D	LM-339	Quad Comparator	
E	XR-14412	FSK Mod/Demod 300 BPS	
F	XR-2120	Filter-Switched Cap	
G	XR-2123	PSK Mod/Demod 1200 BPS	
H	CD-4049	Hex Inverter	
I	CD-4016	Quad B1 - Lateral Switch	
J	CD-4013	Dual D Flip-Flop	
K	CD-4013	Dual D Flip-Flop	
L	XR-1488	Quad Line Driver	
M	XR-1489	Quad Line Receiver	
N	XR-4194	Dual Tracking Regulator	
O	74C193	Synchronous Up/Down Counter	
P	74C193	Synchronous Up/Down Counter	
Q	CD4011	Quad 2 Input Nand Gate	
R	8048/8748	Microprocessor	

C1	82 pF	C17	0.001µF
C2	0.033µF	C18	4.7 µF
C3	0.022µF	C19	2.2 µF
C4	0.1 µF	C20	4.7 µF
C5	0.033µF	C21	4.7 µF
C6	0.033µF	C22	0.1 µF
C7	0.033µF	C23	0.1 µF
C8	0.033µF	C24	4.7 µF
C9	0.033µF	C25	4.7 µF
C10	0.033µF	C26	4.7 µF
C11	0.1 µF	C27	0.1 µF
C12	0.022µF	C28	0.1 µF
C13	4.7 µF	C29	.22µF
C14	1 µF	C31	20 pF
C15	0.1 µF	C32	20 pF
C16	0.001µF	C33	0.1 µF
		C34	1µF

R1	2.2K	R2	2.2K	R3	2.2K			
R4	2.2K	R5	1.2K	R6	1M			
R7	10K	R8	1.6K	R9	1M			
R10	1K	R11	1K	R12	62K			
R13	100K	R14	47K	R15	62K			
R16	10K	R17	100K	R18	470K			
R19	100K	R20	1K	R21	10K			
R22	62K	R23	47K	R24	100K			
R25	18K	R26	62K	R27	1K			
R28	4.7K	R29	10K	R30	1M			
R31	120K	R32	10K	R33	1K			
R34	68K	R35	600	R36	300			
R37	600	R38	10K	R39	10K			
R40	10K	R41	10K	R42	10K			
R43	39K*	R44	180K*	R45	300**			
R46	39K*	R47	180K*	R48	300**			
R49	39K*	R50	300**	R51	180K			
R52	13K	R53	71.5K	R54	10K			
R55	1K	R56	10K	R57	10K			
R58	10K	R59	1M	R60	10K			
R61	10K	R62	220K	R63	20K			
R64	20K	R65	29K	R66	20K			
RA	1K	RB	1K	RC	1K			

All resistor values are in ohms
* 1% Tolerance
** May Need Fine Tune

CRYSTALS

XTAL1 - 4.032 MHz   M-TRON
XTAL2 - 1.000 MHz   FOX
XTAL3 - 4.608 MHz   X-TRON

TRANSFORMER

T1 - T2220   MICROTRAN

TRANSISTORS

Q1 - 2N4403	Q3 - 2N4401	Q4 - 2N4401

FETs

Q2 - 2N4861

DIODES

D1 - D6 1N914

EXAR Bell 212A Parts List

Bell 212A Modem

This circuit is a complete 300/1200 modem that's compatible with the U.S. Bell 212A standard. Its key circuit is the EXAR XR-2123 PSK modulator/demodulator. Source: EXAR Databook, EXAR Corporation.

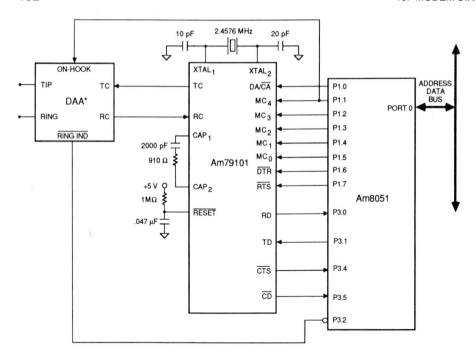

300-bps Full-Duplex Modem
This simple 300-bps modem uses the Am7901 "World Chip," a single-chip frequency shift keying (FSK) modem that's compatible with both Bell and CCITT standards. The Am8051 microcontroller provides a simple interface between the modem chip and the computer. Source: "World Chip FSK Autodial Modem Preliminary Data Sheet," Advanced Micro Devices, Inc.

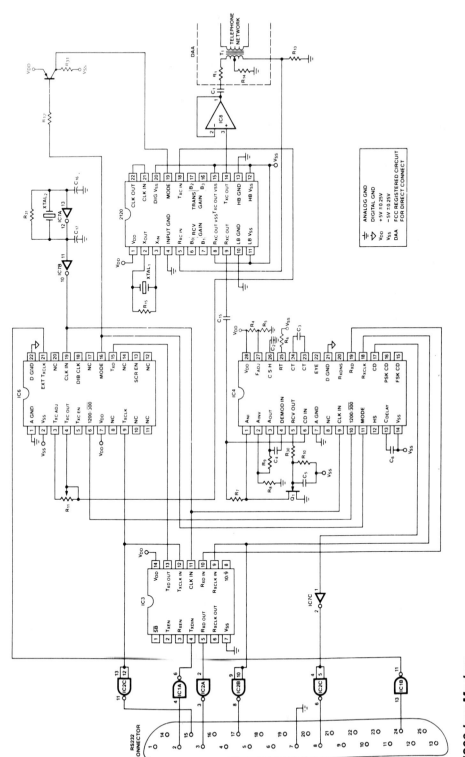

1200-bps Modem

This circuit uses just four chips for a full-featured Bell 212A-type 1200 bps modem. It consists of the XR-2120 PSK modem filter, a switched-capacitor type filter that provides precise filtering and equalization; the XR-2121 PSK/FSK modulator performs all the modulation functions; the XR-2122 demodulator performs demodulation functions; and the XR-2125 data buffer performs asynchronous to synchronous and synchronous to asynchronous conversion. Source: "XR-212AS Modem System," AN-28, EXAR Databook, EXAR Corporation.

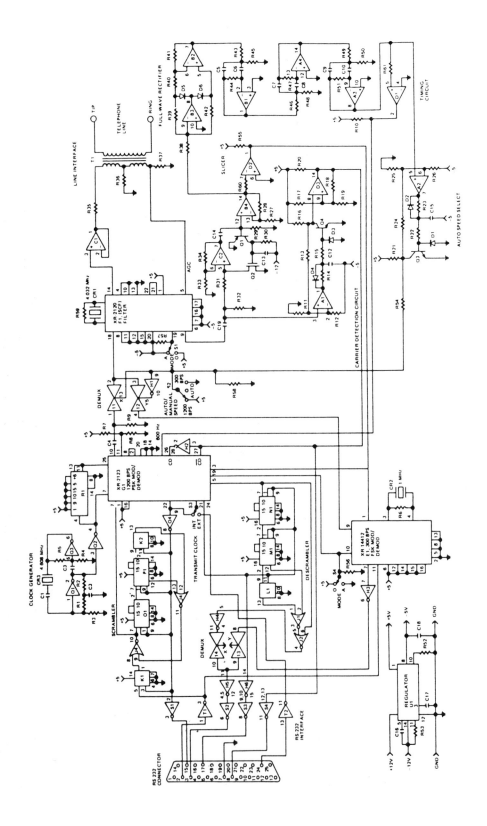

A.	XR-4741 Quad Op Amp					
B.	XR-4741 Quad Op Amp					
C.	XR-1458 Dual Op Amp					
D.	LM-339 Quad Comparator					
E.	XR-14412 FSK Mod/Demod 300 BPS					
F.	XR-2120 Filter-Switched Cap					
G.	XR-2123 PSK Mod/Demod 1200 BPS					
H.	CD-4049 Hex Inverter					
I.	CD-4016 Quad B1-Lateral Switch					
J.	CD-4030 Quad Exclusive-OR Gate					
K.	CD-4013 Dual D Flip-Flop					
L.	CD-4013 Dual D Flip-Flop					
M.	Dual 4 Bit Static Register 4015					
N.	Dual 4 Bit Static Register 4015					
O.	Dual 4 Bit Static Register 4015					
P.	Dual 4 Bit Static Register 4015					
Q.	MM7404 Hex Inverter					
R.	DM74193 Synchronous Up/Down Counter					
S.	XR-1488 Quad Line Driver					
T.	XR-1489 Quad Line Receiver					
U.	XR-4194 Dual Tracking Regulator					

R_1	2.2K	R_2	2.2K	R_3	2.2K	
R_4	2.2K	R_5	1.2K	R_6	1M	
R_7	10K	R_8	10K	R_9	1M	
R_{10}	10K	R_{11}	1K	R_{12}	62K	
R_{13}	100K	R_{14}	47K	R_{15}	62K	
R_{16}	10K	R_{17}	100K	R_{18}	470K	
R_{19}	100K	R_{20}	10K	R_{21}	10K	
R_{22}	62K	R_{23}	47K	R_{24}	100K	
R_{25}	18K	R_{26}	62K	R_{27}	1K	
R_{28}	4.7K	R_{29}	10K	R_{30}	1M	
R_{31}	120K	R_{32}	10K	R_{33}	1K	
R_{34}	68K	R_{35}	600	R_{36}	300	
R_{37}	600	R_{38}	10K	R_{39}	10K	
R_{40}	10K	R_{41}	10K	R_{42}	10K	
R_{43}	39K*	R_{44}	180K*	R_{45}	392*	
R_{46}	39K*	R_{47}	180K*	R_{48}	392*	
R_{49}	39K*	R_{50}	464*	R_{51}	180K*	
R_{52}	13K	R_{53}	71.5K	R_{54}	10K	
R_{55}	10K	R_{56}	10K	R_{57}	10K	
R_{58}	10K	R_{59}	1M	R_{60}	10K	
R_{61}	10K					

All resistor values are in ohms.
* = >1% tolerance.

Crystals
CR1 — 4.032 MHz   MTRON
CR2 — 1,000 MHz   FOX
CR3 — 4.608 MHz   X-TRON

Transformer
T1 — T2220 MICROTRAN

Transistors
Q1 — A854   ROHM
Q3 — C1741   ROHM
Q4 — C1741   ROHM

FETs
Q2 — 2N4861

C_1	82 pF	C_{14}	1 µF	
C_2	.033 µF	C_{15}	.1 µF	
C_3	.022 µF	C_{16}	.001 µF	
C_4	.1 µF	C_{17}	.001 µF	
C_5	.033 µF	C_{18}	4.7 µF	
C_6	.033 µF	C_{19}	2.2 µF	
C_7	.033 µF	C_{20}	4.7 µF	
C_8	.033 µF	C_{21}	4.7 µF	
C_9	.033 µF	C_{22}	.1 µF	
C_{10}	.033 µF	C_{23}	.1 µF	
C_{11}	.1 µF	C_{24}	4.7 µF	
C_{12}	0.22 µF	C_{25}	4.7 µF	
C_{13}	4.7 µF	C_{26}	4.7 µF	

Full-Duplex 300/1200-bps Modem System

This modem operates at either 1200 bps with phase-shift encoding (PSK) or 300 bps with frequency shift keying (FSK). The heart of the system is three LSI ICs. The XR-2120 is a switched-capacitor filter (SCF) that provides precision bandpass filtering at 1200 and 2400 Hz. The XR-2123 performs the 1200-bps PSK modulation/demodulation and the XR-14412 performs the 300-bps FSK modulation/demodulation. The three devices are shown with the necessary external function to perform as a 212A type synchronous modem. Source: "Full-Duplex 1200 bps/300 bps Modem System," AN-25, EXAR Databook, EXAR Corporation.

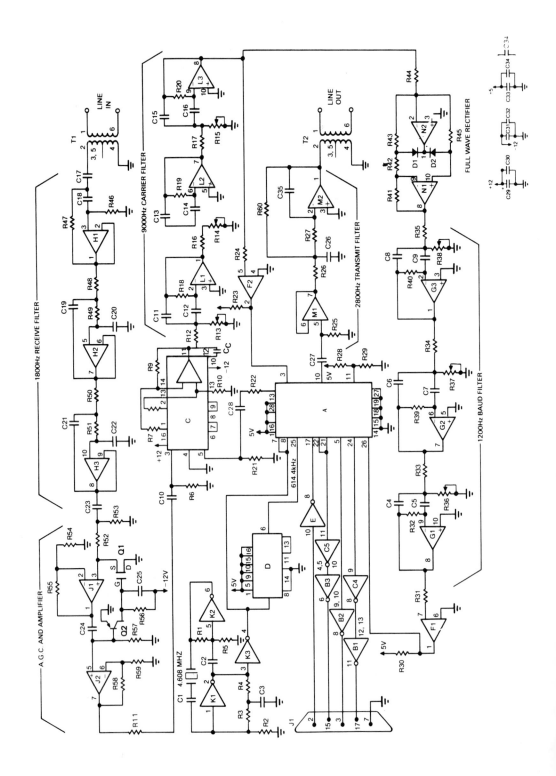

INTEGRATED CIRCUITS			RESISTORS				
A	XR-2123/A	EXAR	R1	1.2K		R21	2K	R41	10K
B	XR-1488	EXAR	R2	2.2K		R22	100K	R42	10K
C	XR-2208	EXAR	R3	2.2K		R23	10K	R43	10K
D	DM-74193	National	R4	2.2K		R24	10K	R44	10K
E	XR-1489	EXAR	R5	1M		R25	1M	R45	10K
F	LM-339-N	Texas Instruments	R6	2K		R26	3.32K	R46	5.76K
G	XR-4741	EXAR	R7	24K		R27	2.2K	R47	2.74K
H	XR-4741	EXAR	R8	24K		R28	1K	R48	2.61K
J	XR-1458	EXAR	R9	50K		R29	1K	R49	75K
K	F-7404	Fairchild	R10	50K		R30	10K	R50	7.87K
L	XR-4741	EXAR	R11	200K		R31	10K	R51	249K
M	XR-1458	EXAR	R12	43.2K		R32	82.2K	R52	120K
N	XR-4741	EXAR	R13	1K POT		R33	29.1K	R53	10K
			R14	1K POT		R34	29.1K	R54	1K
			R15	1K POT		R35	29.1K	R55	68K
			R16	43.2K		R36	500Ω POT	R56	1M
			R17	43.2K		R37	500Ω POT	R57	10K
			R18	109K		R38	500Ω POT	R58	4.7K
			R19	109K		R39	82.2K	R59	1K
			R20	109K		R40	82.2K	R60	6.8K

TRANSISTORS				
Q1	2N4861		Q2	2N4403

TRANSFORMERS		DIODES	
T1	T2220	D1	IN914
T2	T2220	D2	IN914

CONNECTOR	
J1	RS232

CAPACITORS			
C1	82 pf	C19	.01 μf
C2	.0022 μf	C20	.001 μf
C3	.033 μf	C21	.01 μf
C4	.033 μf	C22	100 pf
C5	.033 μf	C23	2.2 μf
C6	.033 μf	C24	2 μf
C7	.033 μf	C25	10 μf
C8	.033 μf	C26	.033 μf
C9	.033 μf	C27	1 μf
C10	.1 μf	C28	.1 μ
C11	.0033 μf	C29	4.7 μf
C12	.0033 μf	C30	.1 μf
C13	.0033 μf	C31	4.7 μf
C14	.0033 μf	C32	.1 μf
C15	.0033 μf	C33	4.7 μf
C16	.0033 μf	C34	.1 μf
C17	.1 μf	C35	.0068 μf
C18	.1 μf		

2400-bps CCITT Modem

2400-bps modems for dial-up telephone lines have become standard in the past few years. This circuit, which is built around the EXAR XR-2123 PSK modulator/demodulator, as shown here is only compatible with the European CCITT V.26 standard. Source: EXAR Databook, EXAR Corporation.

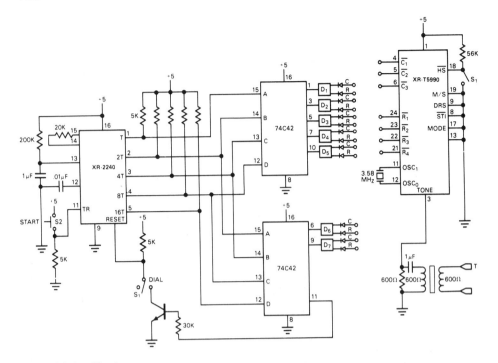

Auto Dialer Modem

Modems that do their own dialing are standard these days, and this simple, 4-chip circuit uses the EXAR XR-T5990 pulse/tone dialer to add that capability to existing nondialing modems. Source: EXAR Databook, EXAR Corporation.

13. MODEM CIRCUITS

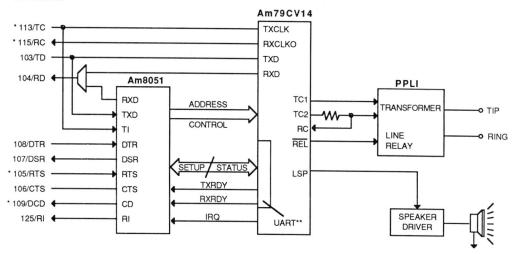

* Not Required by V.25 bis
** 8250 - compatible UART

2400-bps Stand-Alone Intelligent Modem

2400-bps communications is fast becoming a world standard on dial-up telephone lines. The Am79CV14 is a 2400-bps single-chip modem that's compatible with all world communication standards. As shown here, all you need to add to the Am79CV14 is an Am8051 microcontroller and a few parts to construct a full-featured intelligent modem that can be used for both synchronous and asynchronous operation. Source: "World Chip Full-Duplex 2400 bps Quad Modem Data Sheet," Advanced Micro Devices, Inc.

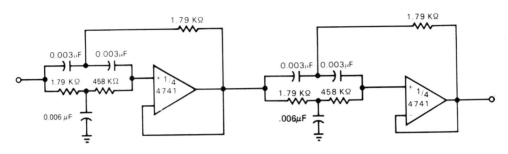

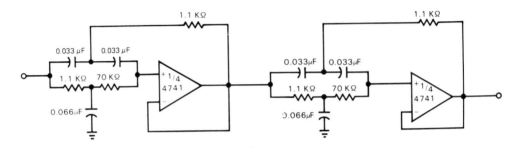

Notch Filters

Notch filters are commonly needed for modem circuitry. Here are two for the common frequencies of 550 and 1800 Hz. Both simple circuits use a minimum of parts and half of an op amp circuit. Both can also be built on the same circuit board using a single op amp. Source: EXAR Databook, EXAR Corporation.

13. MODEM CIRCUITS

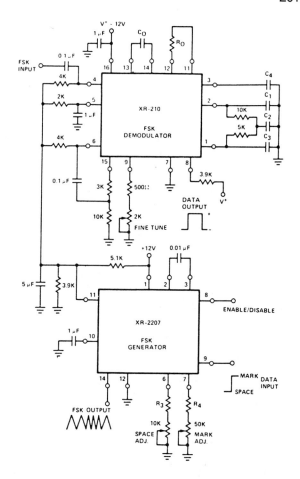

Table 1
Logic Table for Binary Keying Controls

LOGIC LEVEL 8	LOGIC LEVEL 9	SELECTED TIMING PINS	FREQUENCY	DEFINITIONS
0	0	6	f_1	$f_1 = 1/R_3C$, $\Delta f_1 = 1/R_4C$
0	1	6 and 7	$f_1 + \Delta f_1$	$f_2 = 1/R_2C$, $\Delta f_2 = 1/R_1C$
1	0	5	f_2	Logic Levels: 0 = Ground
1	1	4 and 5	$f_2 + \Delta f_2$	1 = >3 V

Full-Duplex FSK Modem

Because of the high integration of formerly discrete parts into monolithic ICs, the job of designing modems is getting simpler and simpler. This full-duplex FSK modem uses just two chips, an XR-210 for FSK demodulation and an XR-2207 VCO for FSK generation. The table shows component values.
Source: EXAR Databook, EXAR Corporation.

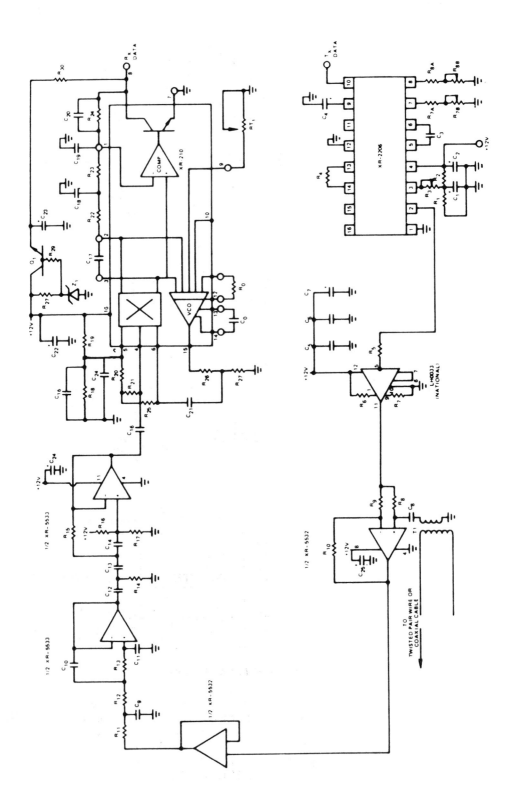

		ANSWER		ORIGINATE
R1-R2	5.1K		5.1K	
R3	50KΩ Pot		50KΩ Pot	
R4	200Ω		200Ω	
R5	51Ω		51Ω	
R6-R7	100Ω		100Ω	
*R8	75Ω		75Ω	
R9-R10	10 KΩ		10 KΩ	
R11-R13	1 KΩ		1 KΩ	
R14	1.1 KΩ		228Ω	
R15	450Ω		90Ω	
R16-R17	16 KΩ		3 KΩ	
R18-R19	5 KΩ		5 KΩ	
R20	2 KΩ		2 KΩ	
R21	4 KΩ		4 KΩ	
R22	10 KΩ		10 KΩ	
R23	5 KΩ		5 KΩ	
R24	249 KΩ		249 KΩ	
R25	4 KΩ		4 KΩ	
R26	3 KΩ		3 KΩ	
R27	10 KΩ		10 KΩ	
R28	5 KΩ		5 KΩ	
R29	562Ω		562Ω	
R30	1.3 KΩ		1.3 KΩ	
RO	2.4 KΩ		7.4 KΩ	
RT2	1 KΩ Pot		1 KΩ Pot	
R7A	1.4 KΩ		562Ω	
R7B	1 KΩ Pot		1 KΩ Pot	
R8A	750Ω		3.3 KΩ	
R8B	1 KΩ Pot		1 KΩ Pot	
RT1	50Ω		100Ω	
C1		47 μf		47 μf
C2		4.7 μf		4.7 μf
C3		.001 μf		.001 μf
C4		4.7 μf		4.7 μf
C5-C6		.1 μf		.1 μf
C7		4.7 μf		4.7 μf
C8		1 μf		1 μf
C9		738 pf		317 pf
C10		1800 pf		807 pf
C11		107 pf		46 pf
C12-C14		1000 pf		1000 pf
C15		.22 μf		.22 μf
C16		1 μf		1 μf
C17		300 pf		300 pf
C18		150 pf		150 pf
C19		106 pf		106 pf
C20		10 pf		10 pf
C21		.1 μf		.1 μf
C22-C25		4.7 μf		4.7 μf
Q1		2N2222A		2N2222A
T1		PE-5760**		PE-5760**
Z1		1N5232		1N5232
IC 1		XR-2206		XR-2206
IC 2		LH0033†		LH0033†
IC 3		XR-5532		XR-5532
IC 4		XR-5533		XR-5533
IC 5		XR-210		XR-210
*150Ω				
J1-J2		JUMPER WIRE		JUMPER WIRE

*Twisted Pair Wire
**Pulse Engineering
‡National

High-Speed FSK Modem

As the need for transmitting large amounts of data increases, there are some applications that require higher speeds than standard telephone lines allow. This high-speed, full-duplex FSK modem uses an XR-2206 as a modulator and an XR-210 as a demodulator. It's capable of transmitting data at up to 100 kilobaud over twisted pair or coaxial cable. The modulator converts the input data to two discrete frequencies corresponding to 1's and 0's and is sent over the line. The line hybrid steers these frequencies to the bandpass filter, where any unwanted signals that are the result of the line are removed. Finally, the demodulator, which is a phase-locked loop, locks onto the incoming frequencies and produces a 1's and 0's output. See the original application note for board layout. Source: "High-Speed FSK Modem Design," AN-26, EXAR Databook, EXAR Corporation.

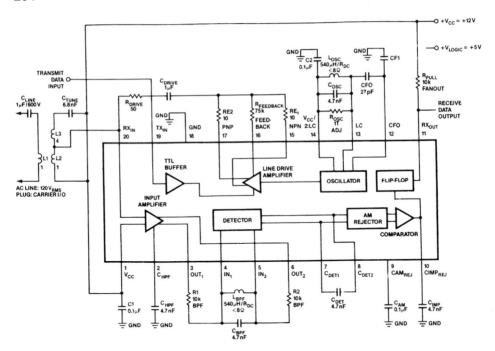

Power Line Modem

The Signetics NE5050 power line modem (PLM), as its name implies, transmits and receives data over a standard AC power line. The AC power line is in general not ideal for data communications because of impulse noise, line impedance modulation, and other factors. The PLM is designed to overcome these problems. It uses two forms of modulation: carrier on/off ASK (amplitude-shift keying) and noncoherent FSK (frequency shift keying). The circuit shown here is just one example of the chip's application. See the original application note for a detailed discussion of the chip's use, as well as important information about electrical shock hazards. Source: "NE5050 Power Line Modem Application Board Cookbook," AN1951, — Linear Data Manual, Volume 1: Communications, Signetics Corporation.

Chapter 14
Optoelectronics Circuits

LED Driver
Photodiode Amplifiers
Light Amplifier
Infrared Remote-Control System
Sensitive Photodiode Amplifier
Photo Diode Detector
High-Speed Photodetector
Balanced Pyroelectric Infrared Detector
100-dB Range Logarithmic Photodiode Amplifier

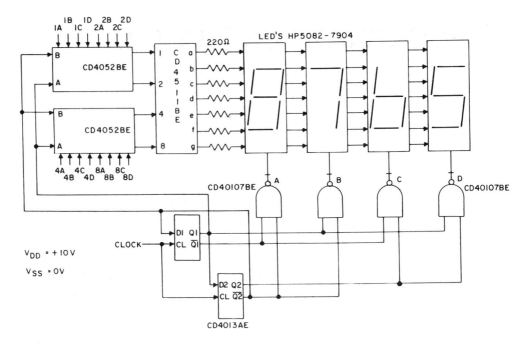

LED Driver

Light-emitting diodes are one of the most commonly used devices today. This simple LED driver is an astable circuit with a single load of the LED with a 150-ohm current limiting resistor. The NAND buffer, as well as driving the load, forms part of the astable circuit, with one of its inputs used as an enable. When this input is low, the LED is off. The other half of the astable oscillator utilizes a two-input NOR gate (the RCA CD4001AE dual NAND buffer), one input of which is used as an inhibit. With the timing components shown, the astable frequency is approximately 4 Hz. Source: D.J. Blanford and G.L. Gimber, "Applications of CD40107BE COS/MOS Dual NAND Buffer," Digital Integrated Circuits Application Note ICAN-6564, RCA Solid State Division.

Photodiode Amplifiers

Because of their high performance, FET op amps are useful for photodiode amplifiers. Here are two examples, both using the Burr-Brown OPA128LM FET op amp. The first circuit offers extremely high sensitivity, while the second works over a wide temperature range. See the original publication (pages 182–183) for a detailed discussion of photodiode design tips. Source: "The Handbook of Linear IC Applications," Burr-Brown.

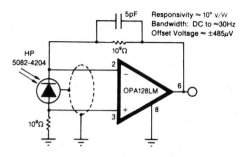

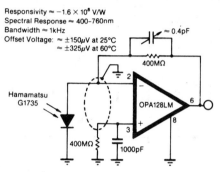

Light Amplifier

This ratiometric light amplifier for absorbance measurement uses the Burr-Brown INA110 very-high-accuracy instrumentation amplifier. In this application, the IC's 4-μsec settling time is used to good advantage. Source: "Integrated Circuits Data Book," Burr-Brown.

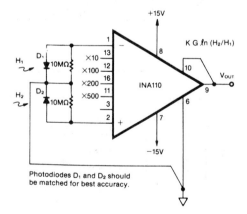

Photodiodes D_1 and D_2 should be matched for best accuracy.

14. OPTOELECTRONICS CIRCUITS

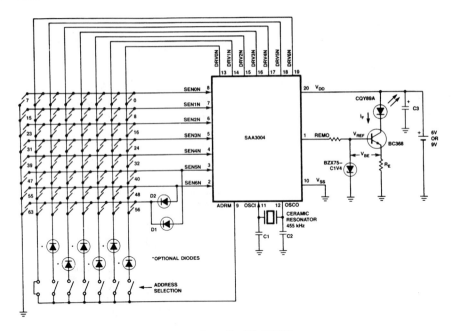

Transmitter With SAA3004

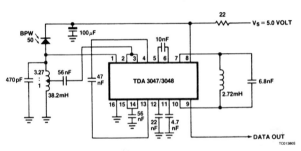

NOTE:
N1 = 3.21
N2 = 1
Q = 16

Narrow-Band Receiver Using TDA3048

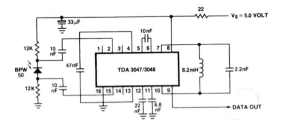

NOTE:
For better sensitivity both 12kΩ resistors may have a higher value.

Wide-Band Receiver With TDA3048

14. OPTOELECTRONICS CIRCUITS

◄ Infrared Remote-Control System

Infrared remote controls are used in a wide range of contemporary applications. The Signetics SAA3004 is a single-chip infrared transmitter for remote control systems. It has a total of 448 commands that are divided into 7 subsystems with 64 commands each. The first circuit here shows a typical application. The two receiver circuits (narrow- and wideband) use the Signetics TDA3048 IR preamplifier. See the original publication for detailed information on using infrared control. Source: Linear Data Manual Volume 3: Video, Signetics Corporation.

Sensitive Photodiode Amplifier

Useful for a wide range of optoelectronic applications, this amplifier uses the Burr-Brown OPA128 ultra-low bias current monolithic operational amplifier. Because of its advanced geometry dielectrically isolated FET (DIFET) inputs, this circuit achieves a performance exceeding even the best hybrid electrometer amplifiers. Source: "Integrated Circuits Data Book," Burr-Brown.

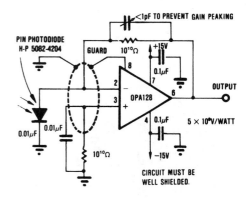

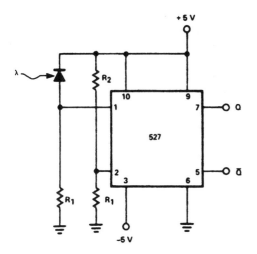

Photo Diode Detector

High speed light sensing is required for a number of applications. This simple circuit uses the inherent speed of the Signetics NE527 voltage comparator. Source: "Applications for the NE521/522/527/529," AN116, Linear Data Manual Volume 2: Industrial, Signetics Corporation.

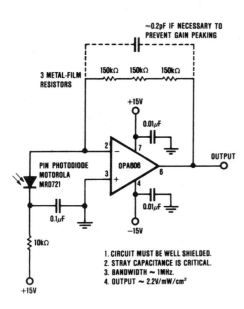

High-Speed Photodetector

This simple circuit is built around Burr-Brown's OPA606 wide-bandwidth monolithic dielectrically isolated FET op amp. Construction notes and specifications are as shown. Source: "Integrated Circuits Data Book," Burr-Brown.

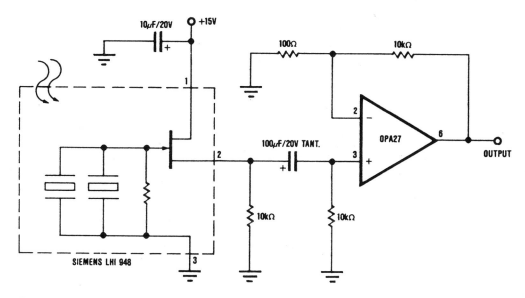

Balanced Pyroelectric Infrared Detector

Infrared detectors are used in a wide variety of contemporary sensing applications. This circuit uses a standard Siemens detector along with Burr-Brown's OPA37A ultra-low noise op amp. Source: "Integrated Circuits Data Book," Burr-Brown.

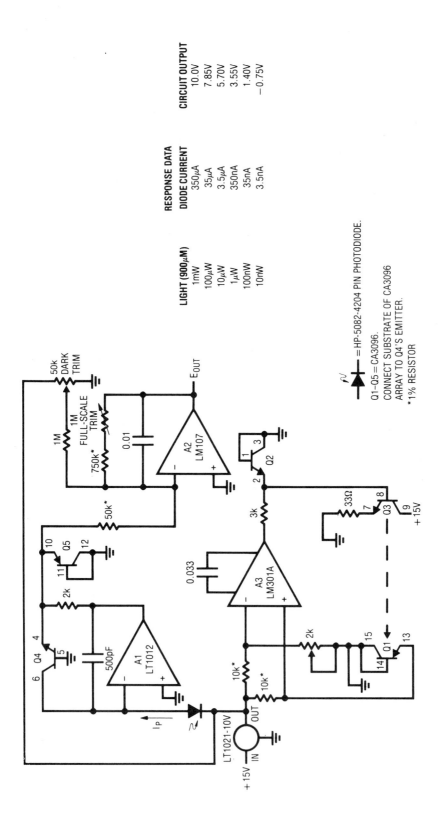

100-dB Range Logarithmic Photodiode Amplifier

PIN photodiodes are used in a wide range of applications. The photodiode in this circuit responds linearly to light intensity over a 100-dB range. Digitizing the diode's linearly amplified output would require an A/D converter with 17 bits of range. But this requirement can be eliminated by logarithmically compressing the diode's output in the signal conditioning circuitry. The circuit's thermal control loop must be set before you can use it. Ground Q3's base and set the 2k potentiometer so A3's negative input voltage is 55 mV above its positive input. This places the servo's setpoint at about 50°C. Unground Q3's base and the array will come to temperature. Next, place the photodiode in a completely dark environment and adjust the "dark trim" so A2's output is 0 V. Finally, apply or electrically simulate 1 mW of light and set the "full-scale" trim for 10 V out. Source: Jim Williams, "Thermal Techniques in Measurement and Control Circuitry," Application Note 5, Linear Technology Corporation.

Chapter 15
Oscillator Circuits

1- to 10-MHz and 1- to 25-MHz Crystal Oscillators
Crystal-Stabilized Relaxation Oscillator
L-C Tuned Oscillator
Temperature-Compensated Crystal Oscillator
Stable RC Oscillator
Reset Stabilized Oscillator
High-Current Oscillator
Crystal-Controlled Oscillator
Synchronized Oscillator
Voltage-Controlled Crystal Oscillator
Voltage-Controlled Oscillator
1-Hz to 1-Mhz Sine Wave Output VCO
Low-Power Temperature Compensated Oscillator
Digitally Programmable PLL
First-Harmonic (Fundamental) Oscillator
A Low-Frequency Precision RC Oscillator
Low Distortion Sinewave Oscillator
Wein Bridge Oscillator
Temperature-Compensated Crystal Oscillator

15. OSCILLATOR CIRCUITS

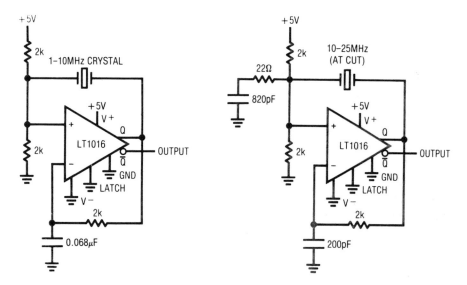

1- to 10-MHz and 1- to 25-MHz Crystal Oscillators

In the first circuit, the LT106 comparator is set up with DC negative feedback. The 2k resistors set the common-mode level at the device's positive input. Without the crystal, the circuit is a very wideband 950 (GHz GBW) unity gain follower biased at 2.5 V. With the crystal inserted, positive feedback occurs and oscillation commences. The second circuit is similar, but supports oscillation frequencies to 25 MHz. Above 10 MHz, AT-cut crystals operate in overtone mode, and oscillation can occur at multiples of the desired frequency. The second circuit ensures proper operation by using a damper network to roll off gain at high frequencies. Source: Jim Williams, "Circuit Techniques for Clock Sources," Application Note 12, Linear Technology Corporation.

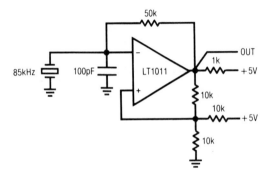

Crystal-Stabilized Relaxation Oscillator

This circuit uses a standard RC-comparator multivibrator circuit with the crystal connected directly across the timing capacitor. Because the free-running frequency of the circuit is close to the crystal's resonance, the crystal "steals" energy from the RC, forcing it to run at the crystal's frequency. In a circuit like this, it's important to ensure that enough current is available to quickly start the crystal resonating while simultaneously maintaining an RC time constant of appropriate frequency. Typically, the free running frequency should be set 5- to 10% above crystal resonance with a resistor feedback value calculated to allow about 100 μa into the capacitor-crystal network. This type of circuit is not recommended for use above a few hundred kHz because of comparator delays. Source: Jim Williams, "Circuit Techniques for Clock Sources," Application Note 12, Linear Technology Corporation.

The specific frequency of operation can be determined by the formula:

$$f = \frac{1}{2\pi\sqrt{LC}}$$ (where C is C_1 and C_2 in series).

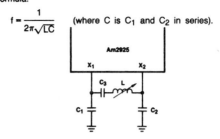

L-C Tuned Oscillator

The Am2925 single-chip, general-purpose, crystal-controlled, clock generator/driver can be operated as a tuned oscillator and will perform as a stable oscillator within the restrictions of the frequency-determining components (the inductor and capacitors) you choose. This classical PI-network with DC loop isolation uses the Am2925 as a DC-biased linear amplifier. In this design the DC bias is necessary and C_3 is included to block the DC path through the inductor. If you use a variable slug-tuned inductor, you can achieve an approximately 2:1 frequency adjustment. The range can be enhanced by switching the two resonant capacitors. Source: "Bipolar Microprocessor Logic and Interface Data Book," Advanced Micro Devices, Inc.

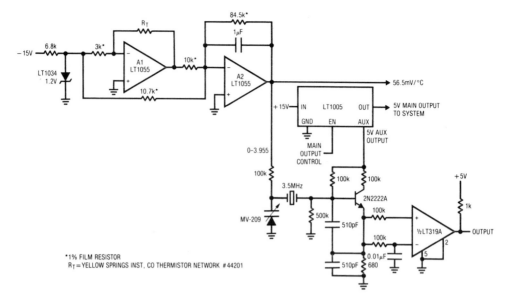

Temperature-Compensated Crystal Oscillator

Although crystal ovens are widely used for minimizing temperature effects on crystal clock frequency, ovens require substantial power and warm-up time. Another approach to offsetting temperature effects, shown in this circuit, is to measure ambient temperature and insert a scaled compensation factor into the crystal clock's frequency-trimming network. This open loop technique relies on matching the clock frequency versus temperature characteristics. This oscillator is a Colpitts type, with a capacitive tapped-tank network. The main regulator output control pin allows the system to be shut down without removing power from the oscillator for overall stability. Source: Jim Williams, "Circuit Techniques for Clock Sources," Application Note 12, Linear Technology Corporation.

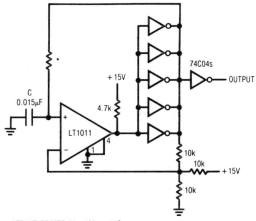

*TRW TYPE MTR-5/ +120ppm/°C
C = 0.015μF = POLYSTYRENE −120ppm/°C ± 30ppm WESCO TYPE 32-P

Stable RC Oscillator

Although pure RC oscillators can't achieve the stability of a synchronized or crystal based approach, they're simple and inexpensive, and are ideal for baud-rate generators or other low-frequency applications. The key to designing a stable RC oscillator is to make output frequency insensitive to drift in as many circuits elements as possible. This RC clock circuit depends primarily on the RC elements for stability, with the RC components chosen for opposing temperature coefficients to further aid stability. This is a standard comparator-multivibrator with parallel CMOS inverters interposed between the comparator output and the feedback resistors. This replaces the relatively large and unstable biplor V_{CE} saturation losses of the LT1011 with the superior ON characteristics of MOS. The paralleling of inverters further reduces errors to insignificant levels. With this arrangement, the charge and discharge time constant of the capacitor is almost totally immune from supply and temperature shifts. Source: Jim Williams, "Circuit Techniques for Clock Sources," Application Note 12, Linear Technology Corporation.

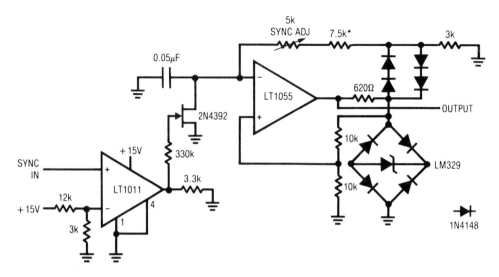

Reset Stablized Oscillator

This circuit's output locks at a higher frequency than the synchronizing input. Circuit operation is the time domain equivalent of a reset stablized DC amplifier, with the LT1055 and its associated components forming a stable oscillator. Source: Jim Williams, "Circuit Techniques for Clock Sources," Application Note 12, Linear Technology Corporation.

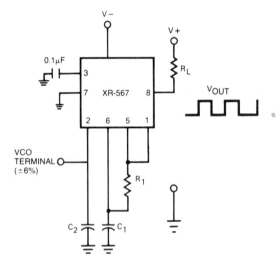

High-Current Oscillator

The EXAR XR-567 monolithic tone decoder has a current-controlled oscillator (CCO) section that provides two basic output waveforms. The square wave is obtained from pin 5, and the exponential ramp from pin 6. As shown here, the oscillator output can be amplified using the output amplifier and high-current logic output available at pin 8. In this way, you can use this circuit to switch 100-mA loads without sacrificing oscillator stability. You can also modulate the oscillator frequency over ±6% by applying a control voltage to pin 2. Source: EXAR Databook, EXAR Corporation.

15. OSCILLATOR CIRCUITS

Crystal-Controlled Oscillator

This simple oscillator circuit uses the industry standard LM139, which consists of four independent precision voltage comparators. Source: Linear Data Manual Volume 2: Industrial, Signetics Corporation.

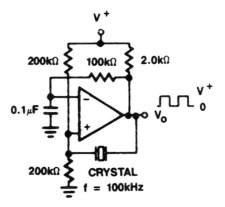

Synchronized Oscillator

This line synchronized clock won't lose lock under noisy line conditions. The basic RC multivibrator is tuned to run free near 60 Hz, but the AC-line-derived synchronizing input forces the oscillator to lock to the line. The circuit derives its noise rejection from the integrator characteristics of the RC network. Source: Jim Williams, "Circuit Techniques for Clock Sources," Application Note 12, Linear Technology Corporation.

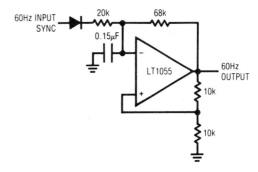

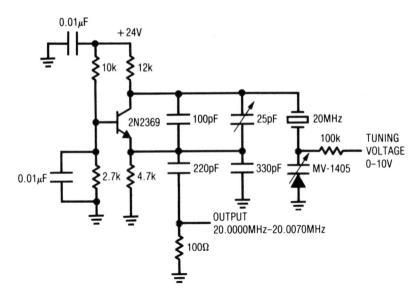

Voltage-Controlled Crystal Oscillator

This VXCO has a clean 20-MHz sine wave output suitable for communications applications. The 25-pF trimmer sets the 20-MHz zero bias frequency. In circuits of this type, it's important to remember that the limit on pulling frequency is set by the crystal's Q, which is high. Circuits such as this one offer pull ranges of several hundred ppm. Larger shifts are possible without losing crystal lock, but the stability of the clock output frequency suffers. Source: Jim Williams, "Circuit Techniques for Clock Sources," Application Note 12, Linear Technology Corporation.

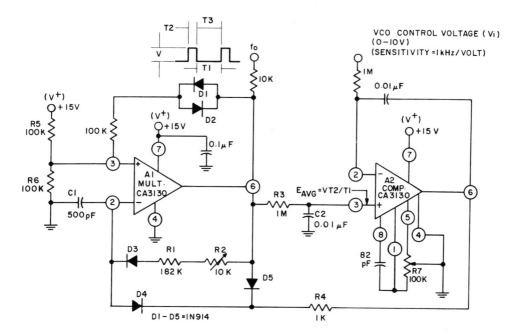

Voltage-Controlled Oscillator

This precision VCO operates with a tracking error on the order of 0.05% and a temperature coefficient of 0.01%/°C. Calibrate this circuit by setting V+ to 15 volts and V_i to 10 volts. Then adjust R_2 until T_2 = 66.6666 microseconds. Then reduce the VCO control voltage (V_i) to 10 millivolts, and adjust the offset nulling potentiometer R_7 until the output frequency is 10 Hz. You should repeat this calibration cycle since secondary adjustments will probably be needed to assure that the VCO operates with optimized linearity up to 10 kHz. Source: H.A. Wittlinger, "Understanding and Using the CA3130, CA3130A, and CA3130B BiMOS Operational Amplifiers," Linear Integrated Circuits Application Note ICAN-6386, RCA Solid State Division.

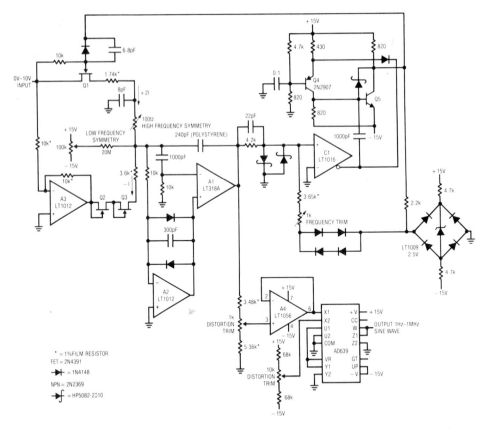

1-Hz to 1-MHz Sine Wave Output Voltage-Controlled Oscillator

Many typical applications such as audio and automatic test equipment require VCOs with a sine wave output. This circuit meets this need, while spanning a 1-Hz to 1-MHz range (120dB, or six decades) for a 0-V to 10-V input. It's also extremely fast, while maintaining 0.25% frequency linearity and 0.40% distortion specifications. To adjust this circuit, put in 10.00 V and trim the 100-ohm pot for a symmetrical triangle at A1. Next, put in 100 µV and trim the 100-k pot for triangle symmetry. Then put in 10.00 V again and trim the 1k frequency trim adjustment for a 100.0-kHz output frequency. Finally, adjust the distortion trim potentiometers for minimum distortion on a distortion analyzer. Source: Jim Williams, "High Speed Comparator Techniques." Application Note 13, Linear Technology Corporation.

15. OSCILLATOR CIRCUITS

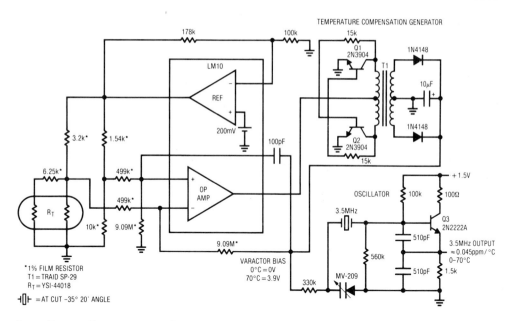

Low-Power Temperature Compensated Crystal Oscillator

Crystal oscillators that run from a 1.5-V supply are relatively easy to build. But if you need good stability over a wide temperature range, things become more difficult. Ovenizing the crystal isn't practical for low-power operation. An alternate method, shown in this circuit, is to provide open loop, frequency correcting bias to the oscillator. The bias value is determined by absolute temperature. In this fashion, the oscillator's thermal drift, which is repeatable, is corrected. Source: Jim Williams, "Circuitry for Single Cell Operation," Application Note 15, Linear Technology Corporation.

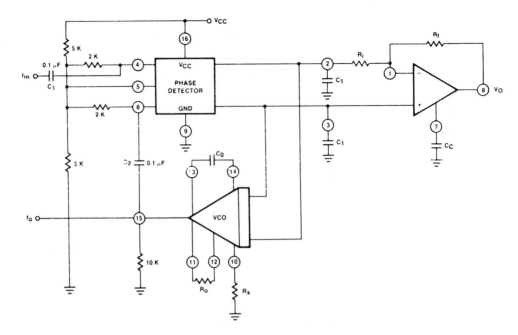

Digitally Programmable PLL

Most phase-locked loops require manual potentiometer adjustment if the center frequency of the circuit is critical. Also, once adjusted, if ambient temperature changes cause the PLL's VCO or center frequency to shift, the potentiometer needs to be readjusted. Because the PLL shown in this circuit is digitally controlled, you can use a microprocessor or other digital circuitry to tune the VCO whenever it's necessary. The circuit uses the XR-215 monolithic PLL together with the XR-9201 D/A converter, which provides the tuning function. As shown here, the PLL has a center frequency of 20 kHz. See the original application note for detailed information on components needed for your desired frequency. Source: "Digitally-Programmable Phase-Locked Loop," AN-24, EXAR Databook, EXAR Corporation.

First-Harmonic (Fundamental) Oscillator

This typical first harmonic oscillator uses the Am2925 single-chip, general-purpose, crystal-controlled, clock generator/driver. The crystal load is comprised of two 68-pF capacitors in series. This 34-pF approximates the standard 32-pF load. If you need a closer match, you should replace one of the capacitors with a parallel combination of a fixed capacitor and trimmer. It's good practice to ground the case of the crystal to eliminate stray pickup. You should also keep all connections as short as possible. Note that at fundamental frequencies below 5 MHz, it's possible for the oscillator to operate at the third harmonic. Source: "Bipolar Microprocessor Logic and Interface Data Book," Advanced Micro Devices, Inc.

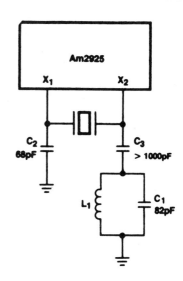

A Low-Frequency Precision RC Oscillator

This simple low frequency RC oscillator uses a medium speed comparator with hysteresis and feedback as timing elements. C1 charges and discharges to $2V+/3$ and $V+/3$ respectively. Because of this, the frequency of oscillation is at theoretically independent from power supply voltages. Source: Nello Sevastopoulos, "Application Considerations for an Instrumentation Lowpass Filter," Application Note 20, Linear Technology Corporation.

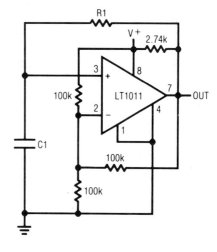

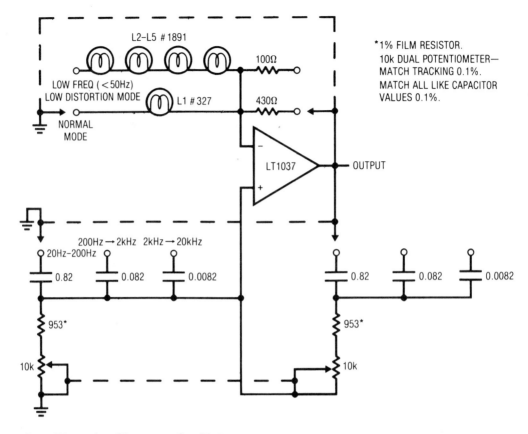

Low Distortion Sinewave Oscillator

This modern adaptation of a classic circuit uses the positive temperature coefficient of lamp filaments. In any oscillator, it's necessary to control the gain as well as the phase shift at the frequency of interest. Too little gain prevents oscillation; too much causes saturation limiting. This circuit uses a variable Wein Bridge to provide frequency tuning from 20 Hz to 20 kHz. Gain control comes from the positive temperature coefficient of the lamp. When power is applied, the lamp is at a low resistance value, gain is high and oscillation amplitude builds. As the amplitude builds, the lamp current increases, heating occurs, and the resistance goes up. This causes a reduction in amplifier game and the circuit finds a stable operating point. Source: Jim Williams, "Thermal Techniques in Measurement and Control Circuitry," Application Note 5, Linear Technology Corporation.

15. OSCILLATOR CIRCUITS

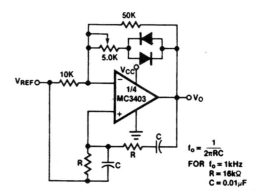

Wein Bridge Oscillator

This contemporary variant on a classic circuit uses the Signetics MC3403 quad op amp with differential inputs. In single-supply applications, this circuit has several advantages over standard op amps. It can operate from supply voltages as low as 3.0 V and as high as 6 V. Also, the common-mode input range includes the negative supply, therefore eliminating the necessity for external biasing components in most applications. Source: "Applications for the MC3402," AN160, Linear Data Manual Volume 2: Industrial, Signetics Corporation.

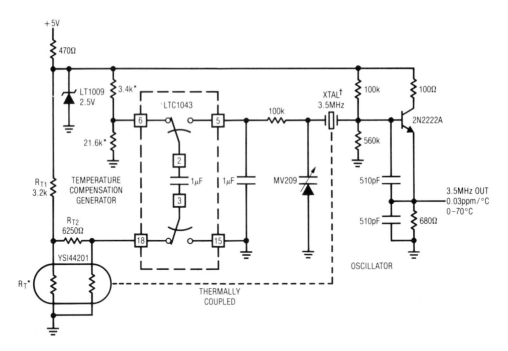

Temperature-Compensated Crystal Oscillator

This circuit uses the LTC1043 to differentiate between a temperature sensing network and a DC reference. The single-ended output biases a varactor-tuned crystal oscillator to compensate for drift. The varactor-crystal network has high DC impedance, thus eliminating the need for an LTC1043 output amplifier. Source: Jim Williams, "Applications for a Switched-Capacitor Instrumentation Building Block," Application Note 3, Linear Technology Corporation.

Chapter 16
Power Supply Circuits

Basic Power Supply
TTL Power Supply Monitor
Programmable Voltage/Current Source
Regulated Negative Voltage Converter
Current Monitor
Two Rectifiers
Bridge Amplifier Load Current Monitor
High-Efficiency Rectifier Circuit
Current-Limited 1-amp Regulator
5-V Regulator
5-V Regulator with Shutdown
Dual Output Regulator
Regulator with Output Voltage Monitor
RMS Voltage Regulator
Switching Regulator
Micropower Post-Regulated Switching Regulator
High-Current Switching Regulator
Switching Preregulated Linear Regulator
Low-Power Switching Regulator
Fully Isolated –48 V to 5-V Regulator
Inductorless Switching Regulator
Single Inductor, Dual-Polarity Regulator
Dual Tracking Voltage Regulator
Current Regulator
Negative-Voltage Regulator
High-Voltage Regulator
Switching Regulator
High-Temperature +15-Voltage Regulator
7.5-A Variable Regulator
Adjustable Regulator
High-Efficiency Regulator

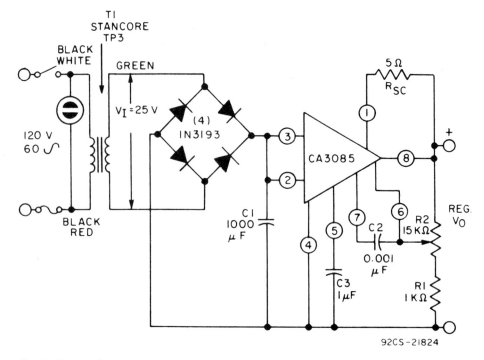

Basic Power Supply

This simple power supply uses the RCA CA3085 voltage regulator. The AC supply voltage is stepped down by T1, full-wave rectified by the diode bridge circuit, and smoothed by the large electrolytic capacitor C1 to provide unregulated DC to the CA3085. Frequency compensation of the error amplifier is provided by capacitor C2. Capacitor C3 bypasses residual noise in the reference voltage source, and decreases the incremental noise voltage in the regulator circuit output. You can adjust the output voltage of this circuit from 1.8 to 20 V DC by varying R2. When you use this circuit to provide high output currents at low output voltages, you have to be careful to avoid excessive IC dissipation. You can do this by increasing the primary-to-secondary transformer ratio, or by using a dropping resistor between the rectifier and the regulator. Source: A. C. N. Sheng and L. R. Avery, "Applications of the CA3085-Series Monolithic IC Voltage Regulators," Application Note ICAN-6157, RCA Solid State.

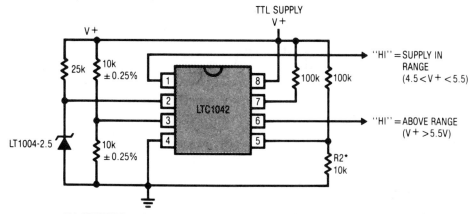

ALL RESISTORS ±5% UNLESS OTHERWISE NOTED.
*SUPPLY TOLERANCE EQUALS R2 IN kΩ. I.E., 10k = ±10%.

TTL Power Supply Monitor

This simple circuit monitors a power supply in situations where it's critical that the supply voltage doesn't exceed specifications. It uses the LTC1042 window comparator, which has extremely low power consumption. Source: "Linear Databook Supplement," Linear Technology Corporation.

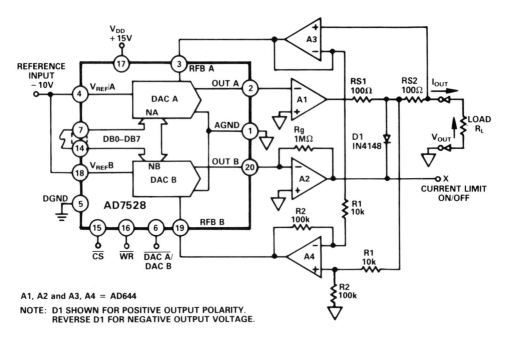

A1, A2 and A3, A4 = AD644

NOTE: D1 SHOWN FOR POSITIVE OUTPUT POLARITY. REVERSE D1 FOR NEGATIVE OUTPUT VOLTAGE.

Programmable Voltage/Current Source

This circuit is a unipolar V/I source that uses Analog Devices' AD7528 dual 8-bit CMOS digital-to-analog (DAC) converter. It requires a negative reference for a positive output voltage. Source: Paul Toomey and Bill Hunt, "AD7528 Dual 8-Bit CMOS DAC Application Note," Application Note E757-15-1/83, Analog Devices.

16. POWER SUPPLY CIRCUITS

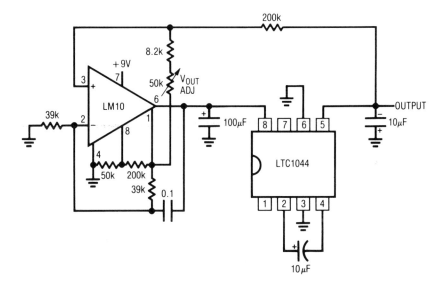

Regulated Negative Voltage Converter

Generating negative voltage is a very common design requirement. But to get the best regulation, you need to decrease output impedance. This circuit encloses a regulator within an op amp's feedback loop for improved regulation. Source: Jim Williams, "Power Conditioning Techniques for Batteries," Application Note 8, Linear Technology Corporation.

Current Monitor

This simple circuit uses a pair of Burr-Brown INA117 precision unity-gain differential amplifiers to monitor current from any AC or DC power supply. Source: "Integrated Circuits Data Book Supplement," Burr-Brown.

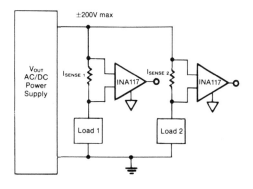

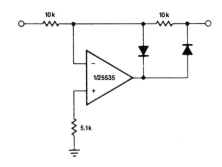

NOTE:
All resistor values are in ohms.

Half-Wave Rectifier

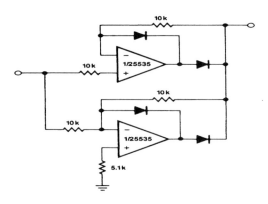

Two Rectifiers

These two rectifiers use the Signetics NE/SE5535 dual high slew rate op amps. The first is a circuit for accurate half-wave rectification of an incoming signal. For positive signals, the gain is 0; for negative signals the gain is -1. You can invert the polarity by reversing both diodes. This circuit provides an accurate output, but the output impedance differs for the two input polarities and buffering may be needed. The second circuit gives you accurate full-wave rectification. The output impedance is low for both input polarities, and the errors are small at all signal levels. Note that the output won't sink heavy currents, except a small amount through the 10 kohm resistors. Therefore, the load applied should be referenced to ground or to a negative voltage. Reversing all diode polarities will reverse the polarity of the output. Source: Linear Data Manual Volume 2: Industrial, Signetics Corporation.

16. POWER SUPPLY CIRCUITS

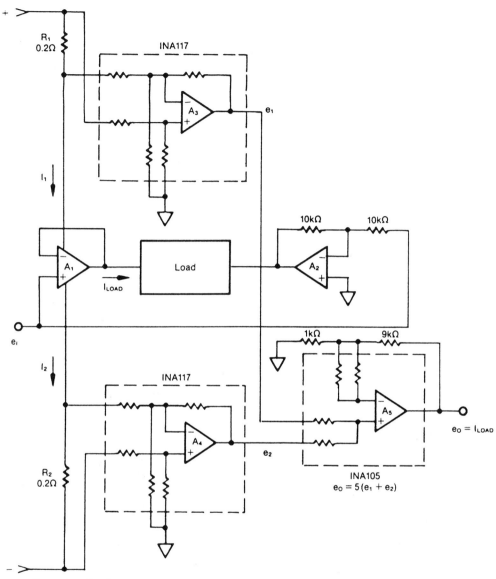

Bridge Amplifier Load Current Monitor
Bridge amplifiers are popular because they double the voltage swing across the load with any given power supply. In this circuit, A1 and A2 form a bridge amplifier driving a load. A1 is connected as a follower and A2 as an inverter. At low frequencies, a sense resistor could be inserted in series with the load and an instrumentation amplifier used to directly monitor the load current. But under high-frequency or transient conditions, CMR errors limit the accuracy of this approach. An alternative approach, as shown here, is to measure the power amplifier supply currents. A Burr-Brown INA117 differential amplifier (A3 and A4) monitors A1 supply currents I1 and I2 across sense resistors R1 and R2. Source: "Integrated Circuits Data Book Supplement," Burr-Brown.

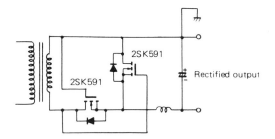

High-Efficiency Rectifier Circuit

Although using dual power MOS FETs improves the efficiency of power supply circuits, the resulting power dissipation of the secondary rectifier circuit becomes significant. This requires a reduction in the power dissipation. This typical high-efficiency rectifier circuit does this while providing a negative power source. No auxiliary secondary winding of the transformer is required. Source: "Power MOS FET Data Book," NEC Corporation

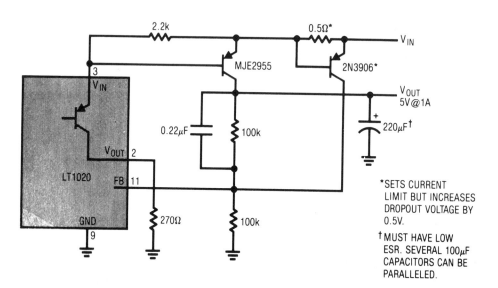

Current-Limited 1-amp Regulator

This circuit uses the Linear Technology LT1020 micropower regulator and comparator to construct a simple 1-amp regulator with current limiting. Source: "Linear Databook Supplement," Linear Technology Corporation.

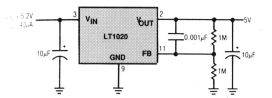

5-V Regulator

This simple 5V regulator uses Linear Technology's LT1020 micropower regulator and comparator. With only 40 μA supply current, the IC can supply over 125 mA of output current. Input voltage can range from 5.2 V to 36 V. Source: "Linear Databook Supplement," Linear Technology Corporation.

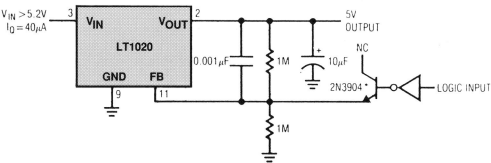

*TRANSISTOR USED BECAUSE OF LOW LEAKAGE CHARACTERISTICS.
TO TURN OFF THE OUTPUT OF THE LT1020
FORCE FB (PIN 11) > 2.5V.

5-V Regulator with Shutdown

This circuit will shut down the output of the LT1020 when it receives a logic input. It's useful for numerous power supply applications. Source: "Linear Databook Supplement," Linear Technology Corporation.

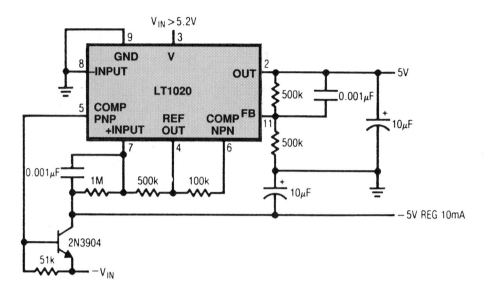

Dual Output Regulator

Here's a voltage regulator that uses the LT1020's dual-output capability to give both +5 V and −5 V output from a single input of from 5.2 to 36 V. Source: "Linear Databook Supplement," Linear Technology Corporation.

16. POWER SUPPLY CIRCUITS

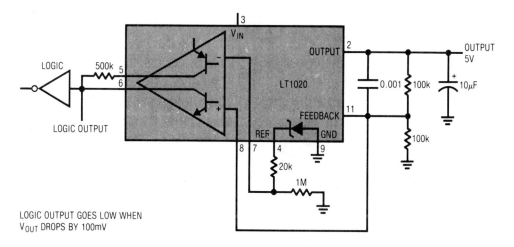

Driving Logic With Dropout Detector

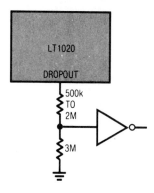

Regulator with Output Voltage Monitor

In many power-supply applications, it's important to know if the supply voltage drops below a particular value. This regulator has a logic output that goes low whenever V_{OUT} drops by 100 mV or more. It uses Linear Technology's LT1020 micropower regulator and comparator, with the dual output comparator used for voltage monitoring in this circuit. Source: "Linear Databook Supplement," Linear Technology Corporation.

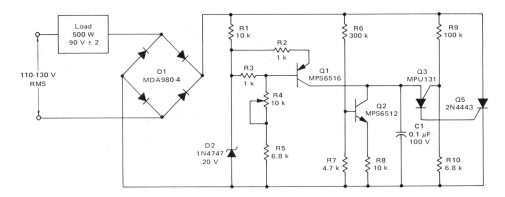

RMS Voltage Regulator

This open-loop RMS voltage regulator will provide 500 watts of power at 90 V RMS with good regulation for an input voltage range of between 110–130 V RMS. To prevent the circuit from latching up at the beginning of each charging cycle, a delay network consisting of Q1 and its associated circuitry is used and prevents the current source from turning on until the trigger voltage has reached a sufficiently high level. This circuit can be operated over a different voltage range by changing resistors R6 and/or R4, which change the charging rate of C1. Source: R. J. Haver and B. C. Shiner, "Theory, Characteristics and Applications of the Programmable Unijunction Transistor," Motorola Semiconductor Products, Application Note AN-527, Copyright of Motorola, Inc. Used by Permission.

Micropower Post-Regulated Switching Regulator ▶

This buck-type switching regulator features a low-loss linear post regulator. Its quiescent current is 40 μA, and it has a output capacity of up to 50 mA. The LT1020 linear regulator provides lower noise than a straight switching approach. It also has internal current limiting and contains an auxiliary comparator that's used to form the switching regulator. Source: Jim Williams, "Micropower Circuits for Signal Conditioning," Application Note 23, Linear Technology Corporation.

16. POWER SUPPLY CIRCUITS

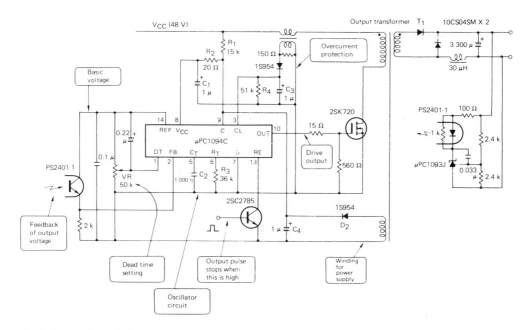

Switching Regulator

This regulator is designed exclusively for power MOS FETs, instead of the more conventional bipolar transistors that are usually used in circuits of this type. The key to this design is using NEC's µPC1094C for switching regulator control. Source: "Power MOS FET Data Book," NEC Corporation.

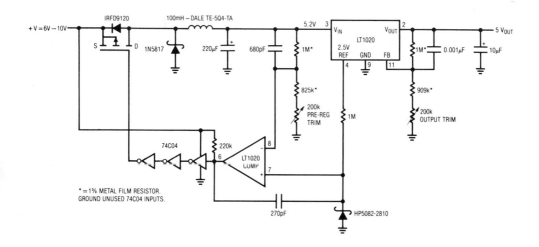

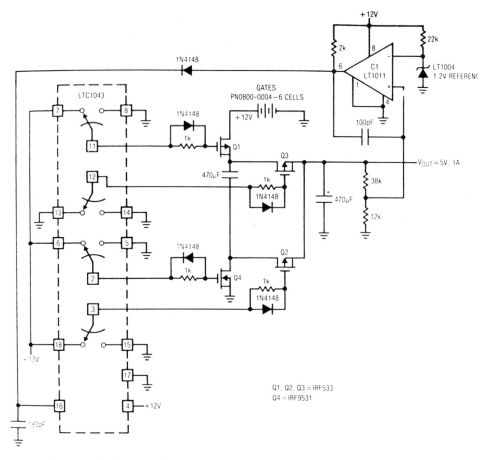

High-Current Switching Regulator

This high-efficiency battery-driven regulator has a 1-A output capacity. In addition, it doesn't require an inductor. This is an unusual feature for a switching regulator operating at this current level. Essentially, this circuit is a large-scale voltage divider that's never allowed to complete a full cycle. It has an overall efficiency of 83%. Source: Jim Williams, "Power Conditioning Techniques for Batteries," Application Note 8, Linear Technology Corporation.

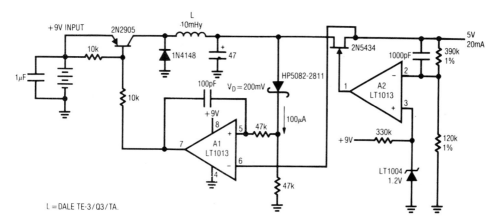

Switching Preregulated Linear Regulator

In some applications, switching induced on a voltage drop regulator may be troublesome. This circuit eliminates any noise by using a low-dropout series regulator at the switching circuit's output. Its overall efficiency is 75%. Source: Jim Williams, "Power Conditioning Techniques for Batteries," Application Note 8, Linear Technology Corporation.

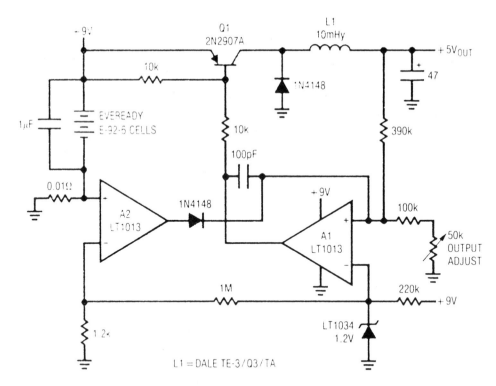

Low-Power Switching Regulator

A low dropout linear regulator is efficient only when its input and output voltages are close. In situations that require substantial voltage drop, you need to use switching techniques to maintain good efficiency. This simple battery-powered switching regulator provides 5 V out from a 9-V source with 80% efficiency and 50-mA output capability. Source: Jim Williams, "Power Conditioning Techniques for Batteries," Application Note 8, Linear Technology Corporation.

16. POWER SUPPLY CIRCUITS

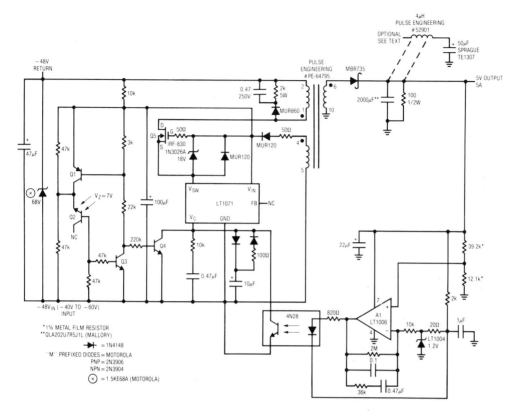

Fully Isolated — 48-V to 5-V Regulator

This telecommunications applications circuit has an output that's fully galvanically isolated from the input, often a requirement. Using the optional filter shown, you can reduce the circuit's broadband output noise, which is about 75 mV p–p. Source: Jim Williams, "Switching Regulators for Poets: A Gentle Guide for the Trepidatious," Application Note 25, Linear Technology Corporation.

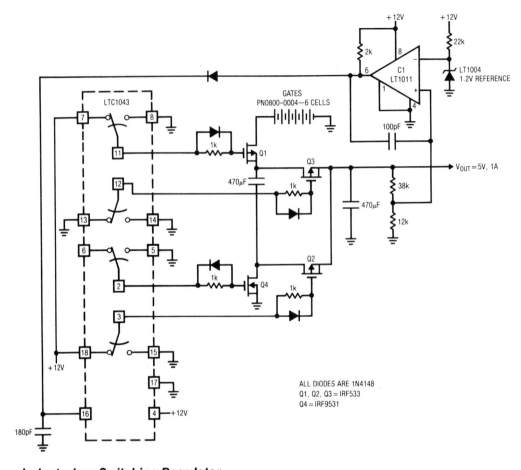

Inductorless Switching Regulator

This high-efficiency battery-driven regulator has a 1 A output capacity. Its lack of an inductor is an unusual feature for a switching regulator that operates at this current level. Essentially, this is a large scale switched-capacitor voltage divider that is never allowed to complete a full cycle. It uses power MOSFETs to easily handle transient current, and has an overall efficiency of 83%. Source: Jim Williams, "Applications for a Switched-Capacitor Instrumentation Building Block," Application Note 3, Linear Technology Corporation.

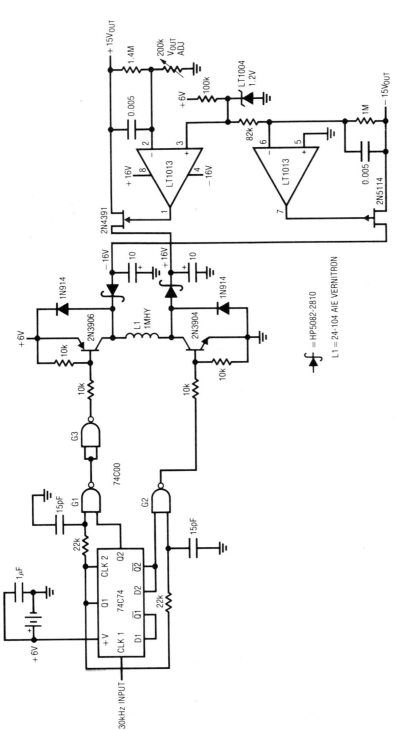

Single Inductor, Dual-Polarity Regulator

This circuit gives you both positive and negative 15-V outputs from a single inductor. It works by alternately determining which end of the inductor is allowed to fly back. The resultant positive and negative peaks are rectified, stored, and regulated to produce a bipolar output. Source: Jim Williams, "Power Conditioning Techniques for Batteries," Application Note 8, Linear Technology Corporation.

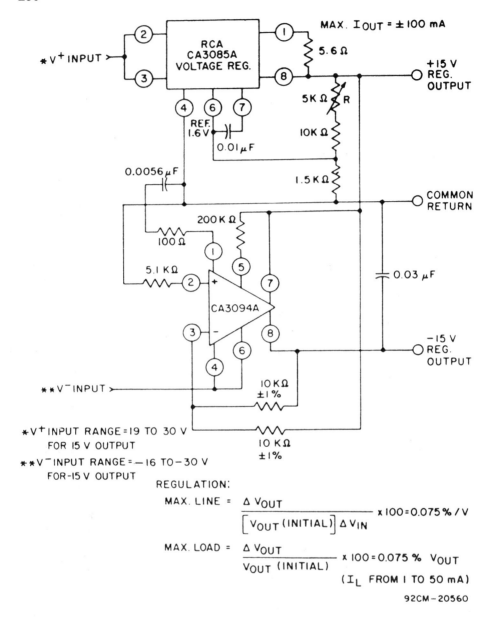

Dual Tracking Voltage Regulator

This circuit uses the RCA CA3085 voltage regulator along with a CA3094A, which is basically an op amp that's capable of supplying 100 milliamps of output current. Resistor R is used as a vernier adjustment of output voltage. You can modify this basic circuit to regulate dissimilar positive and negative voltage (e.g, +15 V DC and −5 V DC) by appropriate selection of resistor ratios in the voltage-divider network. See the original application note for a detailed discussion of this. Source: A.C.N. Sheng and L.R. Avery, "Applications of the CA3085-Series Monolithic IC Voltage Regulators," Application Note ICAN-6157, RCA Solid State.

Current Regulator

This circuit shows how to use a voltage regulator to provide a constant source or sink current. This supply can deliver up to 100 mA. The regulated load current is controlled by R1 because the current flowing through the resistor must establish a voltage difference between terminals 6 and 4 that's equal to the internal reference voltage developed between 5 and 4. Source: A.C.N. Sheng and L.R. Avery, "Applications of the CA3085-Series Monolithic IC Voltage Regulators," Application Note ICAN-6157, RCA Solid State.

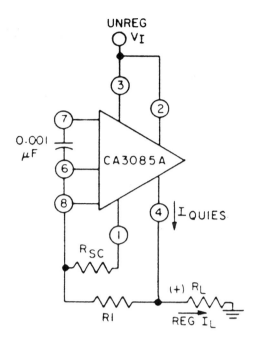

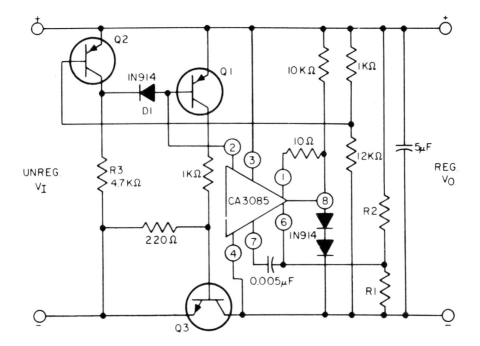

Negative-Voltage Regulator

In this circuit the RCA CA3085 is used as a negative-supply voltage regulator. Transistor Q3 is the series-pass transistor. The CA3085 here is effectively connected across the load side of the regulated system. Diode D1 is used initially in a circuit starter function: transistor Q2 latches D1 out of its starter circuit function so that the CA3085 can assume its role in controlling the pass transistor Q3 by means of Q1. Source: A.C.N. Sheng and L.R. Avery, "Applications of the CA3085-Series Monolithic IC Voltage Regulators," Application Note ICAN-6157, RCA Solid State.

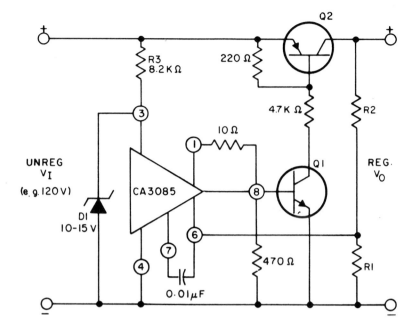

High-Voltage Regulator

This circuit uses the RCA CA3085 voltage regulator as a voltage reference and regulator control device for a high-voltage power supply in which the voltages to be regulated are well above the input voltage ratings of the CA3085 circuits. The external transistors Q1 and Q2 require voltage ratings in excess of the maximum input voltage to be regulated. Source: A.C.N. Sheng and L.R. Avery, "Applications of the CA3085-Series Monolithic IC Voltage Regulators," Application Note ICAN-6157, RCA Solid State.

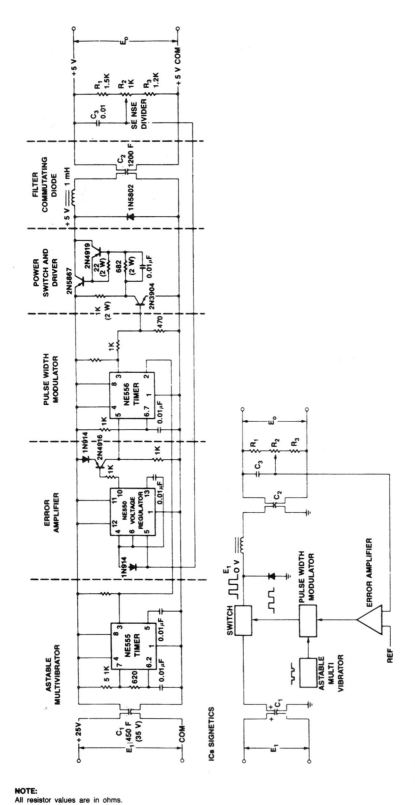

Switching Regulator

Switching power supplies are used in many contemporary digital applications. This switching regulator is built around a pair of standard 555 timers. Source: "NE555 And NE556 Applications," AN170, Linear Data Manual Volume 2: Industrial, Signetics Corporation.

High-Temperature +15-V Voltage Regulator

A regulated source of ±15V is needed in many applications. This circuit fulfills that requirement, while also having the unique ability to operate in temperatures up to 175°C. It accepts +16 V to +30 V at its input and provides +15 V at 20 mA at its output. You can construct a complementary version that provides −15 V by using a 2N1711 transistor. Source: "Integrated Circuits Data Book," Burr-Brown.

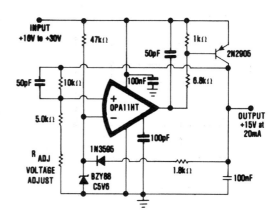

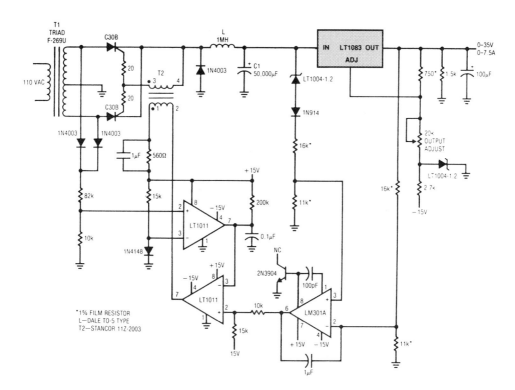

GENERAL PURPOSE REGULATOR WITH SCR PREREGULATOR TO LOWER POWER DISSIPATION. ABOUT 1.7V DIFFERENTIAL IS MAINTAINED ACROSS THE LT1083 INDEPENDENT OF OUTPUT VOLTAGE AND LOAD CURRENT

7.5-A Variable Regulator

This versatile, high-current regulator is based on the Linear Technology LT1083, a high-efficiency positive adjustable regulator. Unlike PNP regulators, where up to 10% of the output current is wasted as quiescent current, the LT1083 quiescent current flows into the load, increasing efficiency. Source: "Linear Databook Supplement," Linear Technology Corporation.

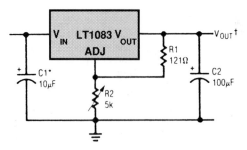

*NEEDED IF DEVICE IS FAR FROM FILTER CAPACITORS

† $V_{OUT} = 1.25V \left(1 + \frac{R2}{R1}\right)$

Adjustable Regulator

This simple regulator produces outputs between 1.2 and 15 V. Its output is calculated using the equation shown. The LT1083 positive adjustable regulator handles loads up to 7.5 A. Source: "Linear Databook Supplement," Linear Technology Corporation.

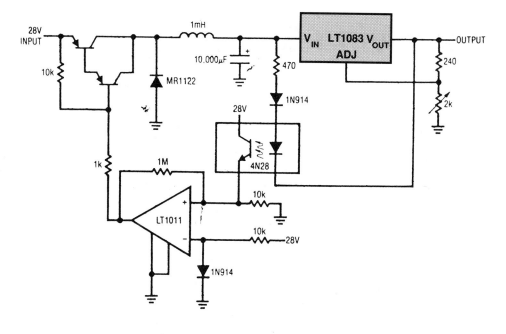

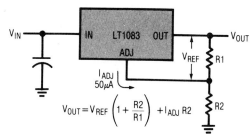

High-Efficiency Regulator

The LT1083 handles loads up to 7.5 A. Here it's used as the heart of a adjustable-output regulator that takes a 28 V input. To calculate the output range for a circuit of this type, see the inset circuit diagram and equation. Source: "Linear Databook Supplement," Linear Technology Corporation.

Chapter 17

Receiving Circuits

SCA Demodulator
FM Tuner
Narrow-Bandwidth FM Demodulator
Clock Regenerator
10.8-MHz FSK Decoder
Narrow-Band FM Demodulator
Linear FM Detector
10.7-MHz IF Amplifier
88- to 108-MHz FM Front End
Balanced Mixer
4–20-mA Current Receiver

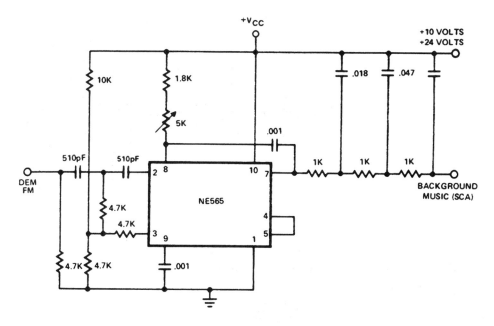

SCA Demodulator

This circuit uses the industry-standard 565 phase-locked loop to demodulate the frequency-modulated subcarrier of the main channel. As shown here, the PLL will recover the SCA (subsidiary carrier authorization) that's mixed with the main carrier of many commercial FM broadcast stations. The SCA signal is a 67-kHz frequency-modulated subcarrier which puts it above the frequency spectrum of normal FM program material. By connecting this circuit to a point between the FM discriminator and the de-emphasis filter of a commercial FM receiver and tuning the receiver to a station that broadcasts an SCA signal, you can receive what's normally commercial-free background music. Source: "Typical Applications with NE565," AN184, — Linear Data Manual, Volume 1: Communications, Signetics Corporation.

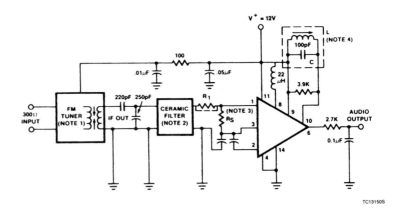

NOTES:
All resistor values are typical and in ohms.
1. Waller 4SN3FIC or equivalent.
2. Murate SFG 10.7mA or equivalent.
3. R_S will affect stability depending on circuit layout. To increase stability R_S is decreased.
   Range of R_S is 330
4. L tunes with 100pF (C) at 10.7MHz Q_O unloaded $\simeq$ 75 (G.I. EX27825 or equivalent).

Performance data at f_O = 98MHz, f_{MOD} = 400Hz, deviation = ± 74kHz.
± 74kHz.
- 3dB limiting sensitivity   2µV (antenna level)
- 20dB quieting sensitivity   1µV (antenna level)
- 30dB quieting sensitivity   1.5µV (antenna level)

FM Tuner

This typical FM tuner with a single-tuned detector coil uses Signetics' CA3089 single-chip FM IF System. This chip is ideal for high-fidelity operations, since distortion is primarily a function of the phase linearity characteristics of the outboard detector coil. Because the CA3089 is a very high-gain device, it's important to give careful consideration to the layout of external components to minimize feedback. The input bypass capacitors should be located close to the input terminals and the values shouldn't be large nor should the capacitance be of the type that might introduce inductive reactance to the circuit. Source: "Linear Data Manual, Volume 1: Communications," Signetics Corporation.

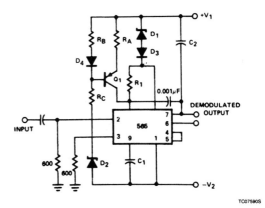

Narrow-Bandwidth FM Demodulator

The 565 is a general-purpose phase-locked loop designed to operate at frequencies below 1 MHz. For applications where both a narrow lock range and a large output voltage swing are required, it's necessary to inject a constant current into pin 8 and increase resistor value, as shown in this circuit. Source: "Circuit Description of the NE565 PLL," AN183, — Linear Data Manual, Volume 1: Communications, Signetics Corporation.

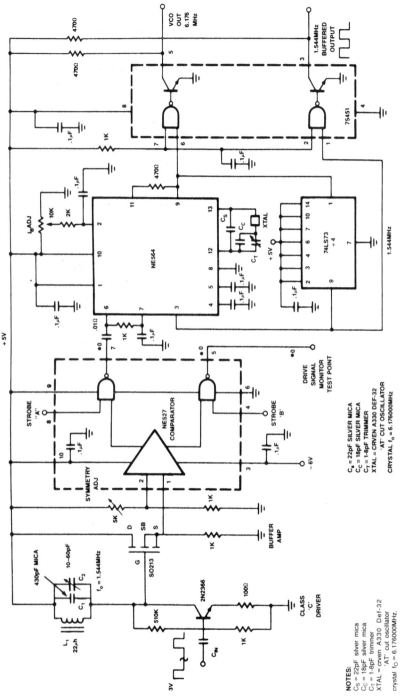

Clock Regenerator

In order to obtain a local clock in multiplexed data transmission systems, a phase and frequency coherent method of signal extraction is required. A master-slave system using a quartz crystal as the primary frequency determining element in a phase-locked loop VCO is used to reproduce a phase coherent clock from an asynchronous data stream. In this circuit, the Signetics NE564 is used to generate the clock reference signal. See the original application note for detailed information on using this circuit. Source: "Clock Regenerator with Crystal-Controlled Phase Locked VCO," AN182, — Linear Data Manual, Volume 1: Communications, Signetics Corporation.

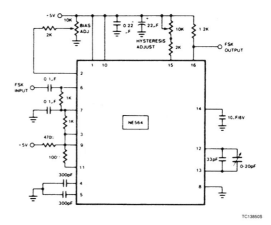

10.8-MHz FSK Decoder

The Signetics NE564 phase-locked loop (PLL) is an industry-standard device that's used in a wide range of applications. It's particularly attractive for FSK demodulation since it contains an internal voltage comparator and a VCO that have TTL compatible inputs and outputs. It can also operate from a single 5-V power supply. Demodulated DC voltages associated with the mark and space frequencies are recovered with a single external capacitor in a DC retriever without utilizing extensive filtering networks. An internal comparator, acting as a Schmitt trigger with an adjustable hysteresis, shapes the demodulated voltages into compatible TTL output levels. The high frequency design of the 564 enables it to demodulate FSK at high data rates in excess of 1 Mbps. Source: "Linear Data Manual, Volume 1: Communications," Signetics Corporation.

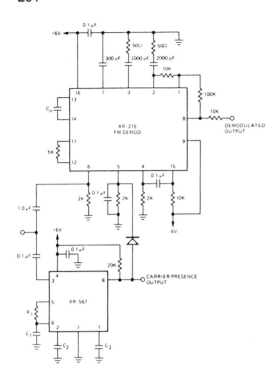

Narrow-Band FM Demodulator

For FM demodulation applications where the bandwidth is less than 10% of the carrier frequency, you can use an EXAR XR-567 monolithic tone decoder to detect the presence of the carrier signal. The output of the XR-567 is used to turn off the FM demodulator when no carrier is present, thus acting as a squelch. In this circuit, an EXAR XR-215 FM demodulator is used because of its wide dynamic range, high signal/noise ratio, and low distortion. The XR-567 will detect the presence of a carrier at frequencies up to 500 kHz. Source: EXAR Databook, EXAR Corporation.

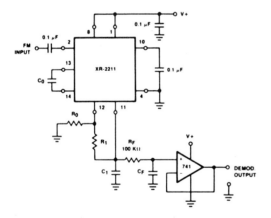

Linear FM Detector

EXAR's XR-2211 FSK demodulator/tone detector can be used as a linear FM detector for a wide range of analog communications and telemetry applications. In this circuit, the demodulated output is taken from the loop phase detector output (Pin 11), through a post-detection filter made up of R_F and C_F, and an external buffer amplifier. See the original publication for details on selecting component values. Source: EXAR Databook, EXAR Corporation.

17. RECEIVING CIRCUITS

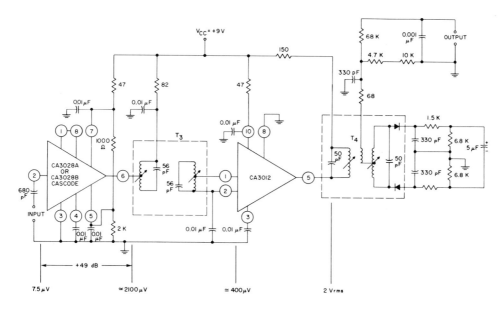

T_3: Interstage transformer TRW #22486 or equiv.
T_4: Ratio detector TRW #22516 or equiv.

Audio Output: 155 mV rms for 7.5 μV ± 75 kHz input 3 dB below knee of transfer characteristic.

10.7-MHz IF Amplifier

This FM IF (Intermediate Frequency) strip uses RCA's CA3028A or CA3028B in a high-gain, high-performance cascode configuration in conjunction with a CA3012 IC wideband amplifier. The CA3012 is used in the last stage because of the high gain of 74 dB input to the 400-ohm-load ratio detector transformer T_4. For −3 dB full limiting, you need an input of approximately 400 microvolts at the base of the CA3012. Source: H. M. Kleinman, "Application of the RCA-CA3028A and CA3028B Integrated-Circuit RF Amplifiers in the HF and VHF Ranges," Application Note ICAN-5337, RCA Electronic Components.

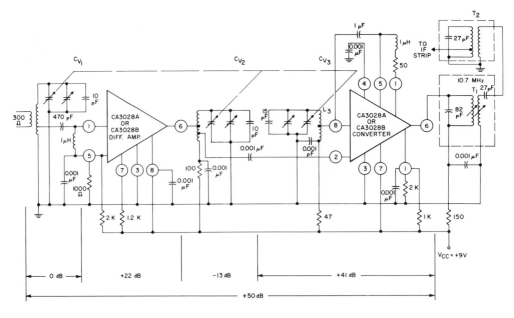

L₁: 3-3/4 T #18 tinned copper wire; winding length 5/16" on 9/32" form; tapped at 1-3/4 T; primary — 2 turns #30 SE.

L₂: 3-3/4 T #18 tinned copper wire; winding length 5/16" on 9/32" form; tapped at 6 2-1/4 T, A 3/4 T.

$C_{V1\text{-}2}$: variable $\Delta C \approx 15$ pF

T_1: Mixer transformer TRW #22484 or equiv.

T_2: Input transformer TRW #22485 or equiv.

L₃: 3-1/2 T #18 tinned copper wire; winding length 5/16" on 9/32" form.

$C_{V1\text{-}3}$: variable, $\Delta C \approx 15$ pF.

88- to 108-MHz FM Front End

This circuit illustrates the use of a single-stage differential amplifier (RCA's CA3028A or CA3028B) as a combination rf amplifier and converter in a front end for a standard FM broadcast receiver. For best noise performance, the differential mode is used and the base of the constant-current source Q_3 is biased for a power gain of 15 dB. Source: H.M. Kleinman, "Application of the RCA-CA3028A and CA3028B Integrated-Circuit RF Amplifiers in the HF and VHF Ranges," Application Note ICAN-5337, RCA Electronic Components.

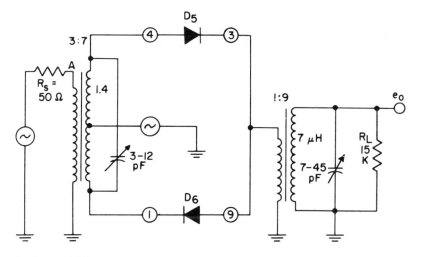

Balanced Mixer

RCA's CA3019 provides four diodes internally connected in a diode–quad arrangement, plus two individual diodes. This IC, although simple, has a wide range of applications including gating, mixing, modulating, and detecting circuits. This circuit uses it as a conventional balanced mixer. The load resistor across the output tuned circuit is selected to provide maximum power output. Source: G.E. Theriault and R.G. Tipping, "Application of the RCA-CA3019 Integrated-Circuit Diode Array." Application Note ICAN-5299, RCA Electronic Components.

4–20 mA Current Receiver

Standard current loops are used in a wide range of control applications. This receiver uses the Burr-Brown INA117 precision unity-gain differential amplifier for low-cost and simplicity. Source: "Integrated Circuits Data Book Supplement," Burr-Brown.

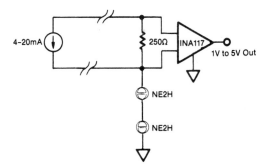

Chapter 18
Signal Circuits

Traffic Flasher
Tone Burst Generator
2.2-Watt Incandescent Lamp Driver
Wide-Range Automatic Gain Control
Low-Droop Positive Peak Detector
High-Speed Peak Detector
Single-Burst Tone Generator
Bandpass Filter for a Multi-Channel Tone Detector

18. SIGNAL CIRCUITS

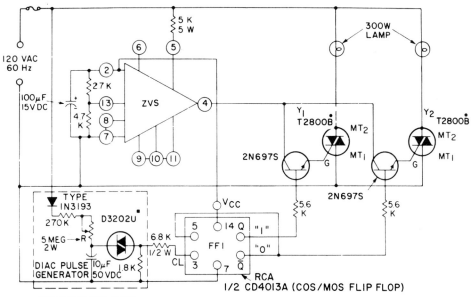

- • FORMERLY RCA 45412
- • FORMERLY RCA 40668

Traffic Flasher

In an application such as switching traffic lamps, it's essential that a triac withstand a current surge of the lamp load on a continuous basis. This surge results from the difference between the cold and hot resistance of the tungsten filament. In this circuit, the lamp loads are switched at zero line voltage. This decreases the required triac surge-current rating, increases the operating lamp life, and eliminates RFI problems. The flashing rate is controlled by potentiometer R, which provides between 10 and 120 flashes per minute. Source: A.C.N. Sheng, G.J. Granieri, J. Yellin, and T. McNulty, "Features and Applications of RCA Integrated Circuit Zero-Voltage Switches," Linear Integrated Circuits Monolithic Silicon, Application Note ICAN-6182, RCA Solid State Division.

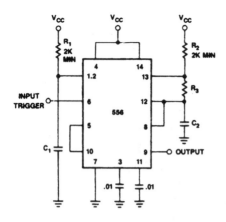

Tone Burst Generator

The 556 dual timer makes an excellent tone burst generator for a variety of applications. All you need to do is connect the first half as a one-shot and the second half as an oscillator. Source: Linear Data Manual Volume 2: Industrial, Signetics Corporation.

NOTE:
All resistor values are in ohms.

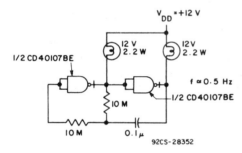

2.2-watt Incandescent Lamp Driver

This circuit uses two NAND buffers, each driving a 2.2-watt, 12-volt incandescent lamp. It's arranged as an astable oscillator with its period of approximately two seconds determined by the external capacitor and resistors. In both this and other similar applications, the load is used as a pull-up from the open-drain output to a power-supply voltage greater than zero and less than or equal to V_{DD}. Source: D.J. Blanford and G.L. Gimber, "Applications of CD40107BE COS/MOS Dual NAND Buffer," Digital Integrated Circuits, Application Note ICAN-6564, RCA Solid State Division.

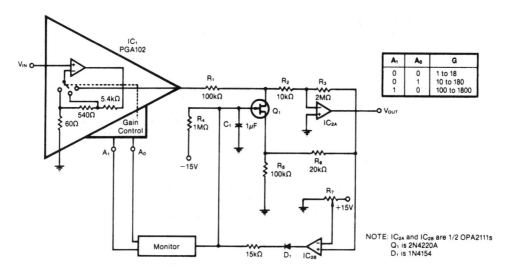

Wide-Range Automatic Gain Control

The Burr-Brown PGA102 monolithic op amp includes its own gain control, thus simplifying the design and layout of signal-acquisition applications. This circuit shows how to add automatic gain control to the chip, for even greater versatility. Unlike many AGC implementations, this signal can accept a wide range of signal amplitudes, up to 1800:1. The monitor circuit here is equivalent to a 2-bit A/D converter that sets the PGA's prescale gain. Source: "The Handbook of Linear IC Applications," Burr-Brown.

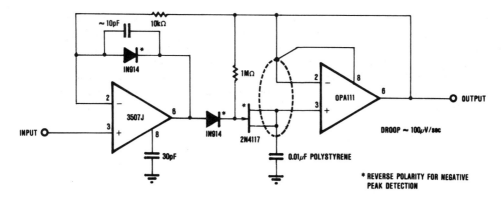

Low-Droop Positive Peak Detector

The heart of this circuit is Burr-Brown's OPA111 precision monolithic dielectrically isolated FET (DIFET) op amp, whose characteristics allow its use in even the most critical instrumentation applications. Source: "Integrated Circuits Data Book," Burr-Brown.

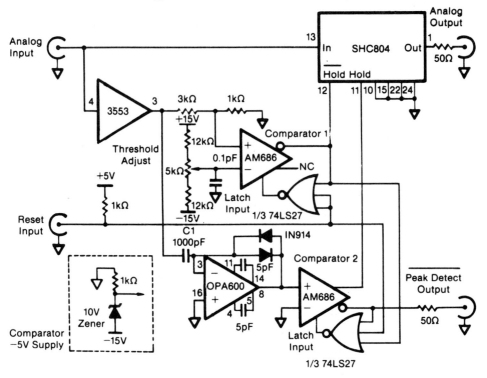

High-Speed Peak Detector

This circuit will capture and hold the peak value of fast-changing analog signals. The occurrence of a peak is detected by the diode-clamped differentiator circuit, which is built around a Burr-Brown OPA600 op amp. When the input voltage reaches a peak and begins to decrease, the current in capacitor C1 reverses, causing the op amp to switch rapidly from its negative to its positive output state. Comparator 2 detects this change and switches the SHC804 into the "HOLD" mode, freezing the peak value. At the same instant, the PEAK DETECT signal goes low, which can be used to initiate a reading of the peak value by an A/D converter. As shown, this circuit is configured to detect positive signal peaks over a ±10-V input range. Note that the output will be inverted due to the -1 gain of the SHC804. If you use careful layout and a good ground plane, this circuit will maintain high accuracy even for very small signals. To use the circuit for negative-going peaks, simply reverse the input polarity on both the comparators. Source: "The Handbook of Linear IC Applications," Burr-Brown.

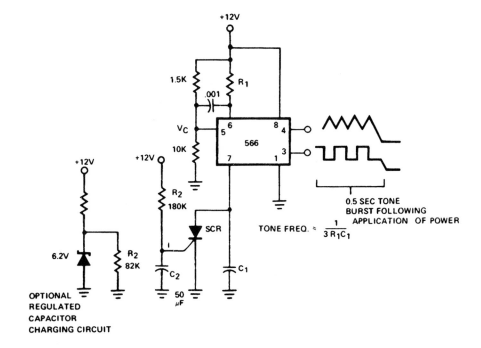

Single-Burst Tone Generator

This tone burst generator supplies a tone for 0.5 second after the power supply is activated. It's primarily intended as a communications network alert signal. The tone is turned off at the SCR, which shunts the timing capacitor C1 charge current when it's activated. The SCR is gated on when C2 charges up to the gate voltage, which occurs in 0.5 second. If you replace the SCR with an NPN transistor, you can switch the tone on and off at the transistor base terminal. Source: "Waveform Generators with the NE566," AN186, — Linear Data Manual, Volume 1: Communications, Signetics Corporation.

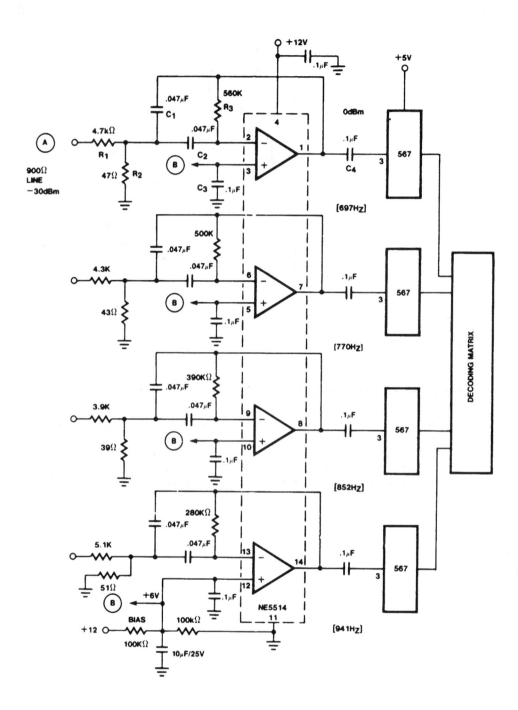

Bandpass Filter for a Multi-Channel Tone Detector

In the design of a multiple-tone signaling system, particularly where signals are transmitted over long lines, noise and adjacent channel interference may be a significant barrier to reliable communications. You can use narrow-band active pre-filters to attain selectivity and gain, and greatly improve the signal-to-noise ratio. Due to its inherent stability, the Signetics NE/SE5514 quad op amp is easy to adapt to such a filter configuration. In addition, its very high input impedance drastically reduces loading to the passive networks and allows for increased Q and large value resistors. This circuit demonstrates multiple feedback filters operating at four of the standard signaling frequencies. You can also add more channels to increase the capacity of the system. Note that the amplifiers are operated from a single +12V supply and are biased to half V_{CC} by a simple resistive divider at point B that connects to all noninverting inputs. Source: "Applications of the NE5514," AN1441, Linear Data Manual Volume 2: Industrial, Signetics Corporation.

Chapter 19
Telephone Circuits

Line-Powered Tone Ringer
Tone Telephone
Basic Telephone Set
Nonisolated 48- to 5-V Regulator
Programmable Multitone Telephone Ringer
Featurephone with Memory
Audio Frequency Sweepers
Bell System 202 Date Encoder and Decoder
Ring Signal Counter
Ring Detector Circuits

19. TELEPHONE CIRCUITS

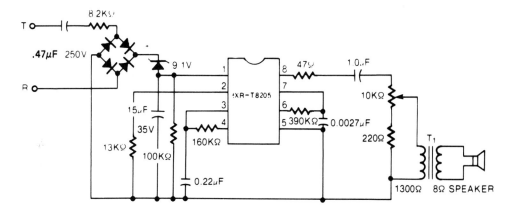

Line-Powered Tone Ringer

EXAR's XR-T8205 tone ringer is primarily intended as an electronic replacement for the "old-fashioned" mechanical telephone bell. Although it can be powered by a separate DC supply, the circuit as shown here is powered directly from the telephone line. Source: EXAR Databook, EXAR Corporation.

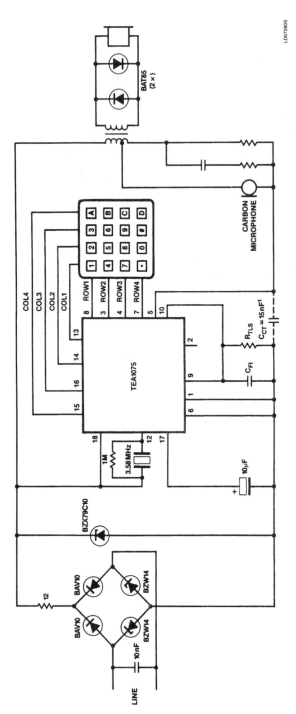

NOTES:
1. C_{CT} connected only if confidence tone is desired.
2. The diagram shows a complete DTMF set including protection.

Tone Telephone

The Signetics TEA1075 is a dual-tone multifrequency (DTMF) generator with line interface for use in telephone sets containing an electronic speed circuit and/or a conventional hybrid transformer. The IC contains a mute switch that handles full line current, which allows two-wire connection between the dial and switch parts. The chip can also be directly controlled from a computer. This circuit shows a typical application, a complete tone telephone. Source: "Linear Data Manual, Volume 1: Communications," Signetics Corporation.

19. TELEPHONE CIRCUITS

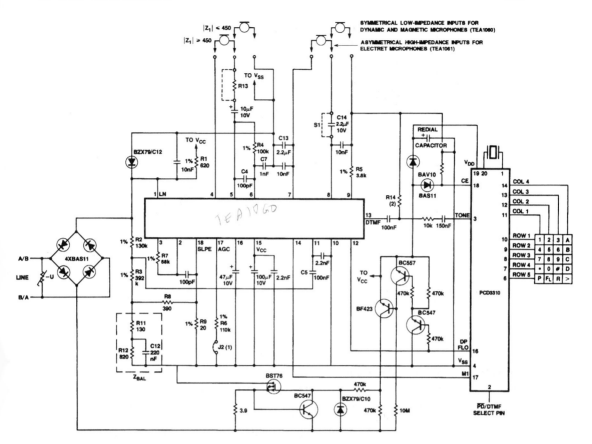

NOTES:
1. Automatic line compenstion obtained by connecting R6 to V_{SS}.
2. The value of resistor R14 is determined by the required level at LN and the DTMF gain of the TEA1060.

Basic Telephone Set

Monolithic ICs are making today's telephones inexpensive and easy to construct, while at the same time providing a host of advanced features. This fully electronic basic telephone uses just two chips. The PCD3310 gives pulse, tone, and automatic redial capabilities. Source: "Linear Data Manual, Volume 1: Communications," Signetics Corporation.

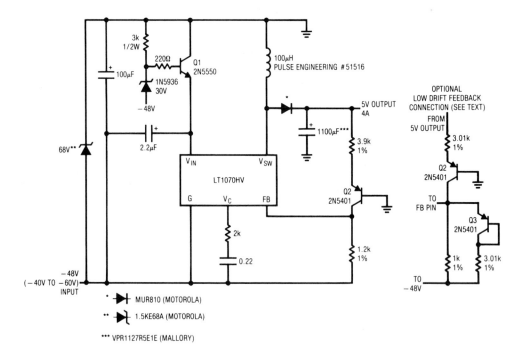

Nonisolated 48- to 5-V Regulator

This circuit is intended mainly for telecommunications applications, where although the raw telecom supply is nominally −48 V, it can vary from −40 to −60V. The circuit's drift is −2 mV/°C, which is normally not objectionable. But you can compensate for it with the optional circuit shown. Source: Jim Williams, "Switching Regulators for Poets: A Gentle Guide for the Trepidatious," Application Note 25, Linear Technology Corporation.

19. TELEPHONE CIRCUITS

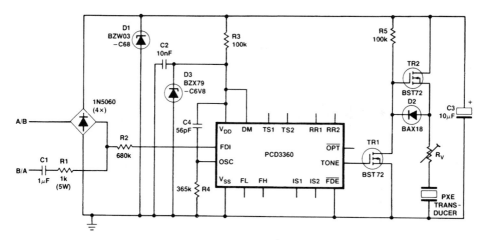

Programmable Multitone Telephone Ringer

If a plain-vanilla telephone bell or even a single-tone electronic ring isn't sufficient enough for your musical tastes, you might try this circuit. It can be programmed to produce 7 basic frequencies, 4 selectable tone sequences, 4 selectable repetition rates, 3-step swell, and a delta modulated output signal that approximates a sinewave. See the original publication for detailed information on setting the chip for your tonal preferences. Source: "Linear Data Manual, Volume 1: Communications," Signetics Corporation.

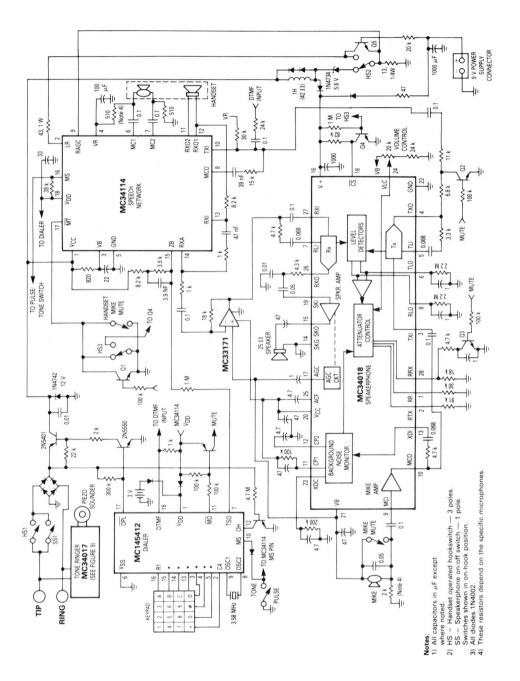

Featurephone with Memory

This circuit shows how to combine the Motorola NC34114 speech network with the MC34018 speakerphone circuit into a "featurephone" that includes the following functions: 10-number memory/pulse dialer, tone ringer, a "privacy" (mike mute) function, and line length compensation for both handset and speakerphone operation. As shown here, the circuit is line-powered except for the 9-V powered speakerphone booster, which ensures adequate volume for speaker use with noisy lines. Note that you should use many of the component values shown as starting points. They may need to be fine tuned for the enclosure and the specific speaker and microphone you use. Source: Dennis Morgan, "A Handsfree Featurephone Design Using the MC34114 Speech Network and the MC34018 Speakerphone ICs," Motorola Semiconductor Application Note AN1002/D, Copyright Motorola, Inc. Used by permission.

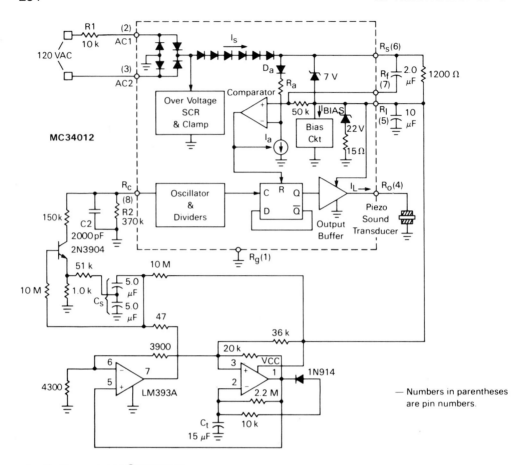

Audio Frequency Sweepers

These two circuits provide a "sweeping" tone in lieu of a standard telephone bell. The two each produce a different effect, though each is similar to video-game effects. The first circuit uses feedback (via the 51-k resistor and the two portions of C8) to generate a different linearity sweep than the simpler second circuit. Source: Dennis Morgan, "A Variety of Uses for the MC34012 and MC34017 Tone Ringers," Motorola Application Note AN933, Copyright of Motorola, Inc. Used by Permission.

19. TELEPHONE CIRCUITS

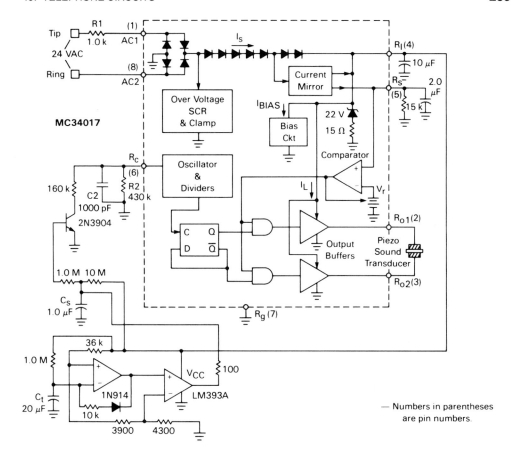

— Numbers in parentheses are pin numbers.

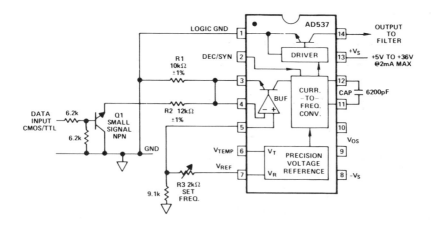

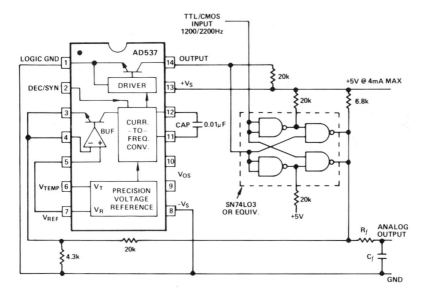

Bell System 202 Data Encoder and Decoder

This first circuit is a FSK (Frequency Shift Keying) encoder with a mark frequency of 1200 Hz and a space frequency of 2000 Hz and has an extremely low current consumption. The inverted connection of Q1 ensures a very small offset error. This circuit only requires one adjustment, the voltage at pin 5 sets both the mark and space frequencies. If you're going to use this circuit over a public telephone line, its square-wave output must be filtered.

For demodulation of a FSK signal, a narrower operating range is required. This first-order loop operates from 800 to 2600 Hz. The center of this range (1700 Hz) is the center of the 1200/2200 type 202 dataset signal. The phase-comparator output is filtered by a simple low-pass circuit. No trimming is required. Source: Doug Grant, "Applications of the AD537 IC Voltage-to-Frequency Converter," Application Note E478-10-8/78, Analog Devices.

19. TELEPHONE CIRCUITS

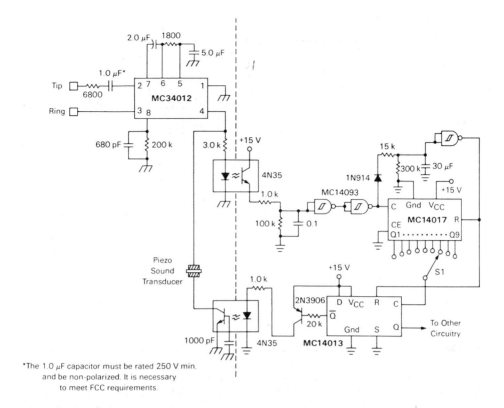

*The 1.0 µF capacitor must be rated 250 V min. and be non-polarized. It is necessary to meet FCC requirements.

Ring Signal Counter

This circuit counts the number of times the ringing voltage is sent to a telephone, and will not allow the piezo transducer to sound until a certain number of ring cycles (selected by S1) have passed. The circuit values are based on a typical ringing cadence of 2 seconds on, 4 seconds off. The outputs of the MC14017 counter (Q1–Q9) become active sequentially with each cycle of the ringing signal. When the selected output becomes active, the MC14013 flip–flop changes state, and the piezo transducer sounds. The Q output of the MC14013 can also be used to activate other circuitry such as an answering machine. After the ringing signal stops, the counters and the flip–flop are reset. Note that the two separate grounds — the telephone line — must be kept separate from the circuit ground. Finally, you can add a volume control by simply connecting a 10k potentiometer in series with the piezo sound transducer. Source: Dennis Morgan, "A Variety of Uses for the MC34012 and MC34017 Tone Ringers," Motorola Application Note AN933, Copyright of Motorola, Inc. Used by permission.

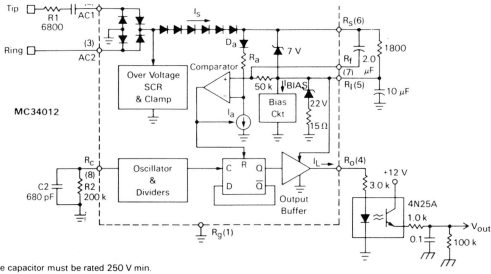

*The capacitor must be rated 250 V min. and be non-polarized. It is necessary to meet FCC requirements.

— Numbers in parentheses are pin numbers.

Ring Detector Circuits

Designed for telephone line use, these two circuits provide an output voltage level to indicate the presence of a ringing signal. Except for using two different ICs, they're essentially similar. The optocoupler provides isolation from the circuit to be controlled, since telephone lines can't be referenced to earth ground. The 1k and 100k resistors, and the 0.1-µF capacitor, filter the square wave output of the IC to provide steady voltage at V_{OUT}. You can use the output to turn on a light, activate an answering machine, etc. These circuits aren't limited to telephones. By changing the value of R1 to 15 kohms for 120 VAC, or to 100 ohms for 24 VAC and deleting the 1.0-µF capacitor, the output will indicate the presence of an input voltage while providing isolation. Isolation can be used to prevent unwanted ground loops, or for safety reasons such as to meet UL requirements. (The 4N25A and 4N35 optocouplers are UL listed.) Source: Dennis Morgan, "A Variety of Uses for the MC34012 and MC34017 Tone Ringers," Motorola Application Note AN933, Copyright of Motorola, Inc. Used by permission.

19. TELEPHONE CIRCUITS

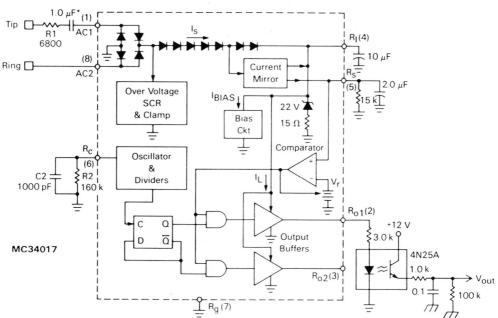

*The capacitor must be rated 250 V min. and be non-polarized. It is necessary to meet FCC requirements.

— Numbers in parentheses are pin numbers.

Chapter 20
Test Circuits

0-5 Amp, 7-30 Volt Laboratory Supply
Pulse Generator
Wide-Range Frequency Synthesizer
Analog Multiplier with 0.01% Accuracy
Computer-Controlled Digitizer
Leakage Current Monitor
Computer-Controlled Digitizer
Voltage Reference
1-nsec Rise Time Pulse Generator
50-MHz Trigger
Active and Passive Signal Combiners
Voltage References
Precision Programmable Voltage Reference
Cable Tester
Low-Frequency Pulse Generator
12-bit Digitally Programmable Frequency Source
Voltage Programmable Pulse Generator
Ultraprecision Variable Voltage Reference
Low-Noise Instrumentation Amplifier
Gain-Ranging Amplifier

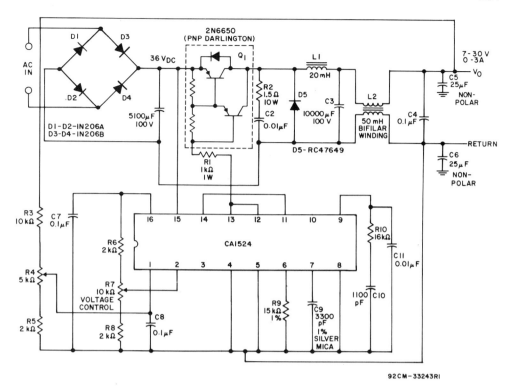

0–5 Amp, 7–30 Volt Laboratory Supply

The CA1524 pulse width modulator IC is used here for a variable output voltage power supply. By connecting the two output transistors in parallel, the duty cycle is doubled. Transistor Q1, an RCA 8203B PNP Darlington transistor, is used as the switching pass element. Its base is driven by the CA1524's outputs. Variability is obtained by first presetting the error amplifier inverting input (terminal 1) to 3.4 volts by appropriate selection of values for the resistor network R3, R4, and R5 in accordance with the maximum output voltage desired. RFI is usually generated with any switching regulator, and R2 and C2 provide a snubber network for the switching current transients of diode D5 to reduce the RFI level. In addition, the output filter network L2 and C4 through C6 provides a bifilar coil that also suppresses the switching noise. Source: Carmine Salerno, "Application of the CA1524 Series Pulse-Width Modulator ICs," Linear Integrated Circuits, Application Note ICAN-6915, RCA Solid State Division.

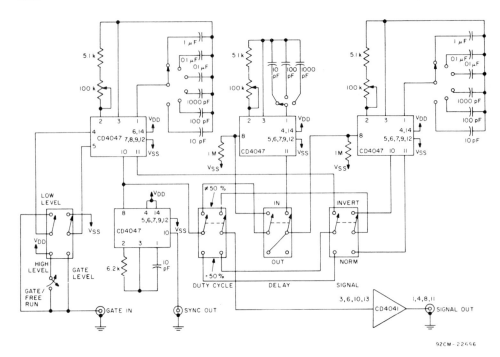

Pulse Generator

Several RCA CD4047A monostable/astable multivibrators can be connected together to produce a general-purpose laboratory pulse generator. This circuit has variable-frequency and pulse-width control, as well as gating and delayed sync capability. Gating can be controlled from a high- or low-level input. This circuit also has automatic 50% duty-cycle capability, as well as normal or inverted output. The signal output is buffered with the CD4041 to allow the pulse generator to drive any required load. This circuit has the advantages of being compact, battery-powered, and COS/MOS compatible. In addition, it's capable of being run from the same power supply as the device under test to assure that the input levels are the same as V_{DD} when the power-supply voltage is varied. Source: J. Paradise, "Using the CD4047A in COS/MOS Timing Applications," Digital Integrated Circuits Application Note ICAN-6230, RCA Solid State Division.

Wide-Range Frequency Synthesizer

This tracking PLL has a tracking range of greater than 100:1 with no harmonic locking problems. The circuit uses a XR-2122 precision phase-locked loop in conjunction with the XR-320 monolithic timer and an XR-084 quad BIFET operational amplifier to form a wide-range PLL with automatic tuning. Source: "Designing Wide-Tracking Phase-Locked Loop Systems," AN-18, EXAR Databook, EXAR Corporation.

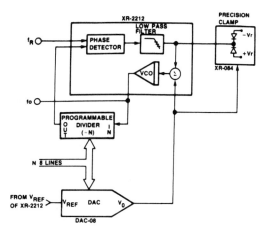

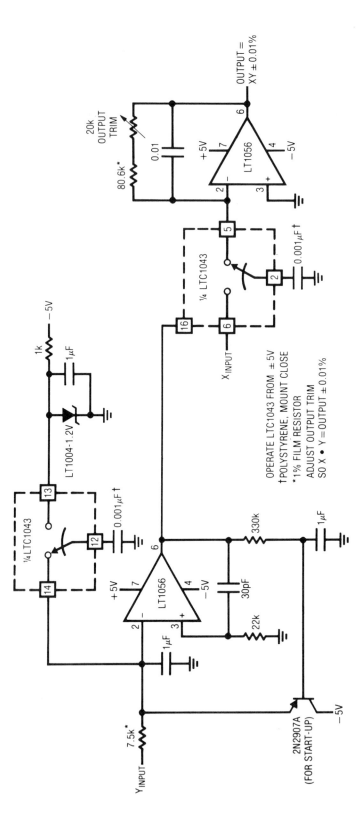

Analog Multiplier with 0.01% Accuracy

This circuit uses a combination of voltage-to-frequency and frequency-to-voltage converters to form a high-precision analog multiplier. The F/V input frequency is locked to the V/F output because the LTC1043's clock is common to both sections. To calibrate this circuit, short the X and Y inputs to 1.7320 V and trim for a 3-V output. Source: Jim Williams, "Applications for a Switched-Capacitor Instrumentation Building Block," Application Note 3, Linear Technology Corporation.

20. TEST CIRCUITS

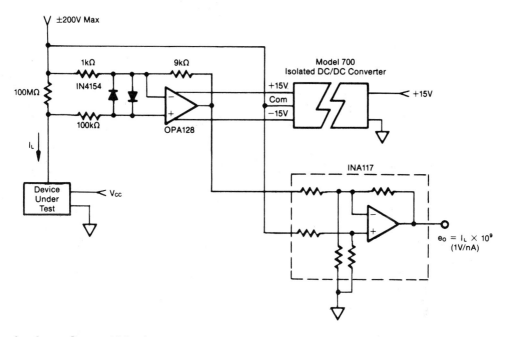

Leakage Current Monitor

In many test conditions a return path isn't independently available. In this case you need to measure leakage current in series with the input. This circuit has a noninverting gain of 10, and an amplifier bias current of less than 75 fA. The diodes and the 100-kohm resistor protect the amplifier from 200-V short-circuit fault conditions. Since common-mode rejection is the ratio of common-mode gain to differential gain, CMR is boosted. In this case the 20-dB gain of the OPA128 operational amplifier is added to the 86-dB CMR of the INA117 differential amplifier for a total CMR of 106-dB minimum. Source: "Integrated Circuits Data Book Supplement," Burr-Brown.

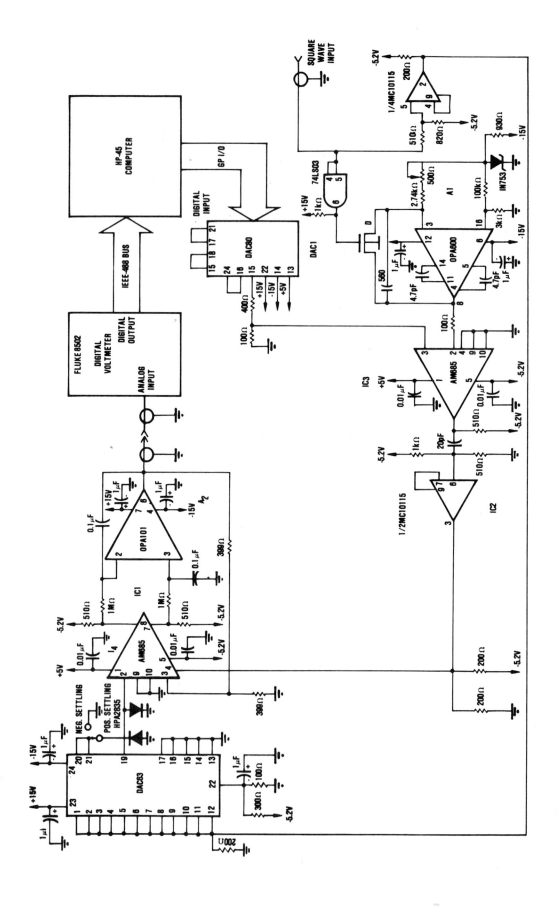

Computer-Controlled Digitizer

This digitizer can be used to record signals for many different measurement requirements. The basic principle behind it is that when a waveform is synchronously sampled, the sampled value of the waveform will correspond to the phase difference between the sample pulse and the input waveform itself. See the original Burr-Brown publication (pages 253–257) for a detailed discussion of the circuit's capabilities. Source: "The Handbook of Linear IC Applications," Burr-Brown.

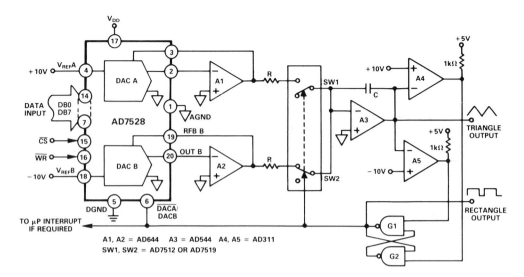

Digitally Programmable Waveform Generator

This circuit is a triangle/rectangle wave generator in which the period of each half cycle can be programmed. It's useful for vector scan CRT displays to generate variable rate sweep signals. A special case occurs when the code in either DAC is zero. Since the integrator input voltages will be zero, the circuit will stop oscillating. You can overcome this limitation by connecting a 10-Mohm resistor from the V_{REF} terminal to the output terminal of each DAC. This provides sufficient bias to keep the circuit oscillating. Source: Paul Toomey and Bill Hunt, "AD7528 Dual 8-Bit CMOS DAC Application Note," Application Note E757-15-1/83, Analog Devices.

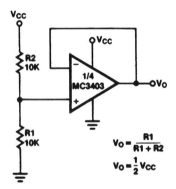

Voltage Reference

This simple reference circuit uses the Signetics MC3403 quad op amp with differential inputs. In single-supply applications, this circuit has several advantages over standard op amps. It can operate from supply voltages as low as 3.0 V and as high as 6 V. Also, the common-mode input range includes the negative supply, therefore eliminating the necessity for external biasing components in most applications. Source: "Applications for the MC3402," AN160, Linear Data Manual Volume 2: Industrial, Signetics Corporation.

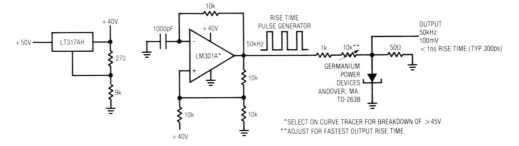

1-nsec Rise Time Pulse Generator

The 10-nsec response of comparators like Linear Technology's LT1016 and associated circuitry challenge even the best test equipment, pushing it to the limits of its capabilities. If you'll be using test equipment with circuits of this type, it's a good idea to verify parameters such as probe and scope rise time and differences in delays between probes and even oscilloscope channels. To do this, you'll need a source of very fast and clean pulses. This circuit uses a tunnel diode to generate a pulse with a rise time of under 1 ns. Source: Jim Williams, "High Speed Comparator Techniques." Application Note 13, Linear Technology Corporation.

50-MHz Trigger

Trigger circuits are often required in counters and other instruments, but designing a fast and stable trigger isn't easy, often requiring a considerable amount of discrete circuitry. This circuit is a simple trigger with 100-mV sensitivity at 50 MHz. To calibrate this circuit, ground the input and adjust the input zero control for 0 V at Q2's drain terminal. Source: Jim Williams, "High Speed Comparator Techniques." Application Note 13, Linear Technology Corporation.

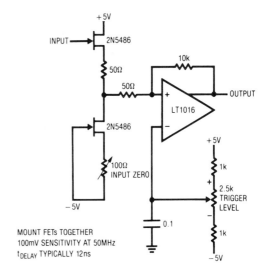

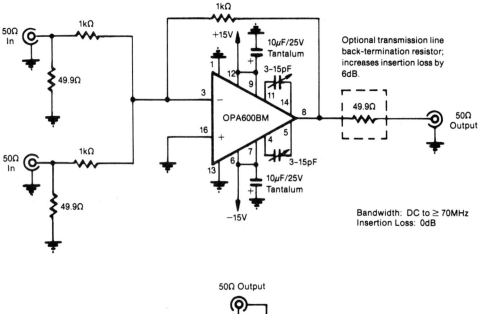

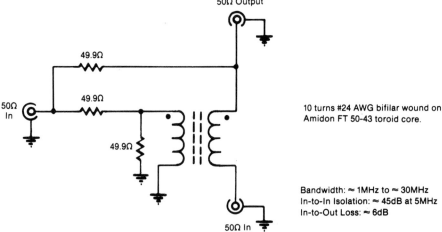

Active and Passive Signal Combiners

In order to prevent intermodulation distortion (IMD), using two separate signal generators for two-tone testing is a common application. These two circuits show two ways of combining the signals. The first is an active summing amplifier that uses a Burr-Brown OPA600 op amp. It provides excellent performance with DC to 5 MHz with harmonic and intermodulation distortion figures typically better than -70 dBC. The second is a passive combiner that has a range from 1 MHz to 30 MHz. The port-to-port isolation will be about 45 dB between signal generators and the input–output insertion loss will be about 6 dB. Source: "Integrated Circuits Data Book Supplement," Burr-Brown.

Voltage References

Here are three variations on voltage reference circuits. All use the Burr-Brown REF101 precision voltage reference. The first gives ±10 V, the second +10 V and +5 V, and the third five different ascending reference voltages. Source: "Integrated Circuits Data Book," Burr-Brown.

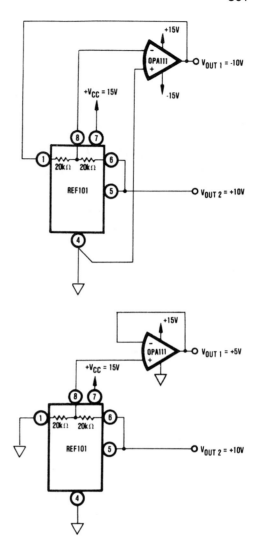

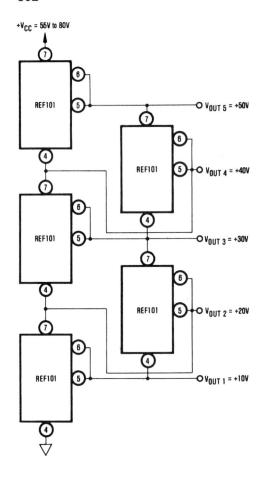

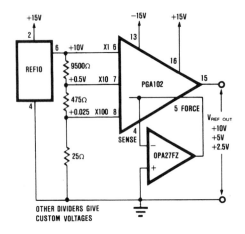

Precision Programmable Voltage Reference

Many testing applications require a precision voltage reference. As shown here, this circuit gives +10-, +5-, and +2.5-V outputs. You can easily modify it for other voltages. See the original publication for detailed information on using the PGA102 digitally controlled op amp. Source: "Integrated Circuits Data Book," Burr-Brown.

20. TEST CIRCUITS

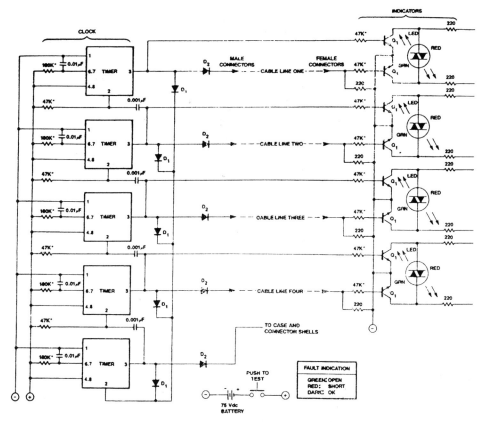

NOTES:
D_1 Silicon Signal Diode
D_2 1-A Silicon Diode
Q_1 200mW NPN Silicon, 2N2926 or ½ 2N2903

Timer Signetics NE555V
LED Monsanto MV5491
*½W Resistors

Cable Tester

This compact tester checks cables for open- or short-circuit conditions. A differential transistor pair at one end of each cable line remains balanced as long as the same clock pulse generated by the timer IC appears at both ends of the line. A clock pulse just at the clock end of the line lights a green LED, and a clock pulse only at the other end lights a red LED. Source: "NE555 And NE556 Applications," AN170, Linear Data Manual Volume 2: Industrial, Signetics Corporation.

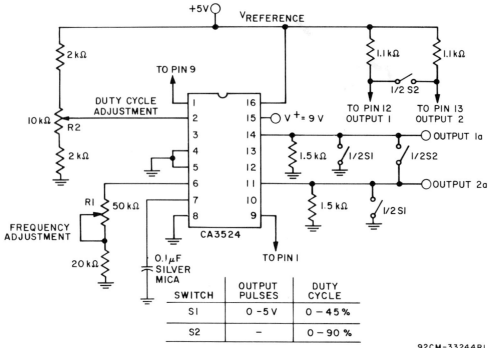

Low-Frequency Pulse Generator

In this circuit, the RCA CA1524 pulse width modulator IC is used to construct a low-frequency pulse generator. Since all components (error amplifier, oscillator, oscillator reference regulator, and output transistor drivers) are on the IC, a regulated 5-V (or 2.5-V) pulse of 0% to 45% (or 0% to 90%) on time is possible over a frequency range of 150 to 500 Hz. Switch S1 is used to go from a 5-V output pulse (switch closed) to a 2.5-V output pulse (switch open) with a duty cycle range of 0% to 45%. Switch S2 allows both output stages to be paralleled for an effective duty cycle of 0% to 90% with the output frequency range from 150 to 500Hz. The frequency is adjusted by R1; R2 controls the duty cycle. Source: Carmine Salerno, "Application of the CA1524 Series Pulse-Width Modulator ICs," Linear Integrated Circuits Application Note ICAN-6915, RCA Solid State Division.

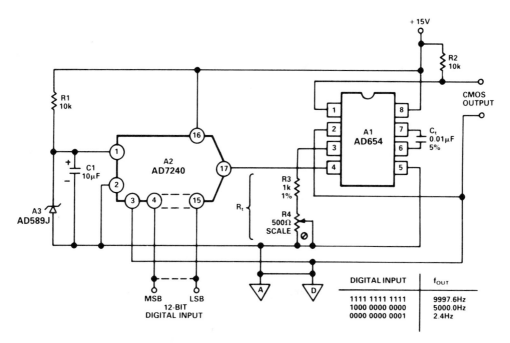

12-bit Digitally Programmable Frequency Source

This circuit uses the AD7240, a 12-bit resolution CMOS D/A converter that's available with differential nonlinearity as low as 1/2 LSB. It's used here with an AD589 1.235-V reference in a voltage output mode. The AD654 is scaled here for a 1-mA FS current, a C_t of 1000 pF, and an FS output of 99976 Hz (with an "all ones" input). Source: Walt Jung, "Operation and Applications of the AD654 IC V-to-F Converter," Application Note E923-25-7/85, Analog Devices.

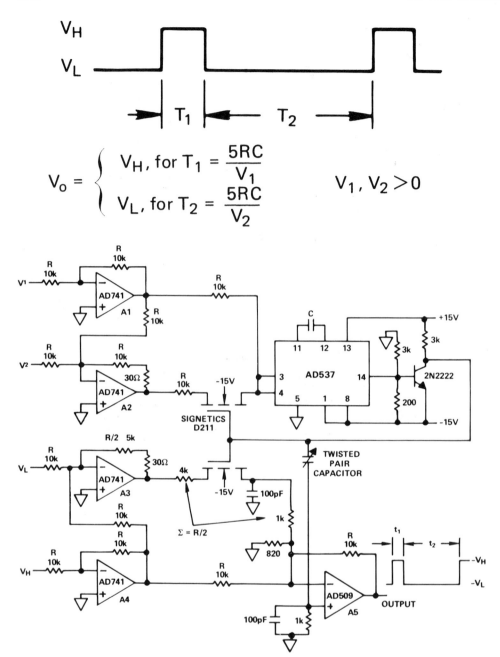

Voltage Programmable Pulse Generator
This circuit produces a pulse train output with continuously variable output frequency, duty cycle, and output levels. It can easily achieve frequency sweeps in excess of 1000 to 1, and even more range is possible by trimming the offsets. Source: Doug Grant, "Applications of the AD537 IC Voltage-to-Frequency Converter," Application Note E478-10-8/78, Analog Devices.

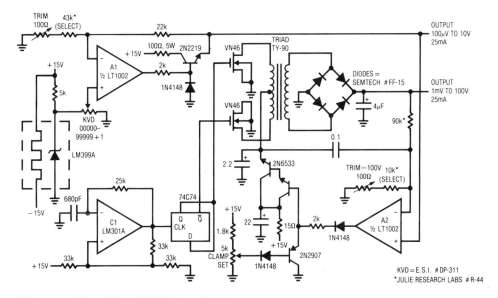

Ultraprecision Variable Voltage Reference

Combining an LT1002 precision op amp with a MOSFET-switched toroid, this voltage reference has a wide dynamic range of outputs. It has two outputs: the low voltage range spans 0 to 10 V and is settable in 100 μV increments. The high voltage range runs from 0 to 100 V in 1-mV steps. To calibrate this circuit, select the 43-ohm value and adjust the 100-ohm trim for a precise 10.000-V output at the low-voltage output. Next, select the 10-kohm value and trim the 100-ohm trimmer in the high voltage divider string. Source: Jim Williams, "Applications of New Precision Op Amps," Application Note 6, Linear Technology Corporation.

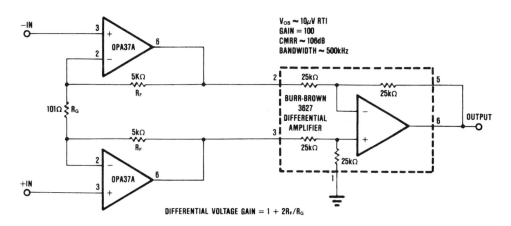

Low-Noise Instrumentation Amplifier

In instrumentation applications, low-noise is often a primary design consideration. The Burr-Brown OPA37A op amps used in this circuit are ultra-low noise. Note the grounded shielding necessary for the 3627 differential amplifier. Source: "Integrated Circuits Data Book," Burr-Brown.

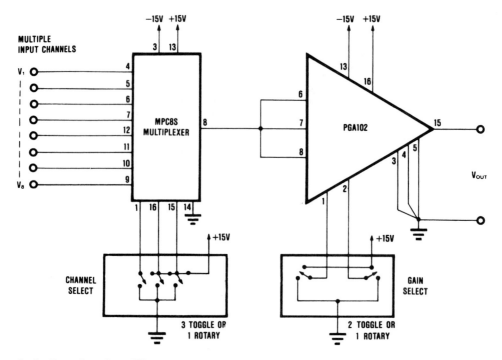

Gain-Ranging Amplifier
This manually controlled, gain-ranging amplifier is ideal for use with portable test equipment. The key to its versatility is the Burr-Brown PGA102 programmable-gain, digitally controlled, fast-settling operational amplifier. Source: "Integrated Circuits Data Book," Burr-Brown.

Chapter 21
Transmitting Circuits

80-watt, 175-MHz Transmitter
Balanced Modulator
Frequency Doublers
Amplitude Modulator
Broadband 160-watt Linear Amplifier
Three VHF Amplifiers
4- to 20-mA Current Loop Transmitter

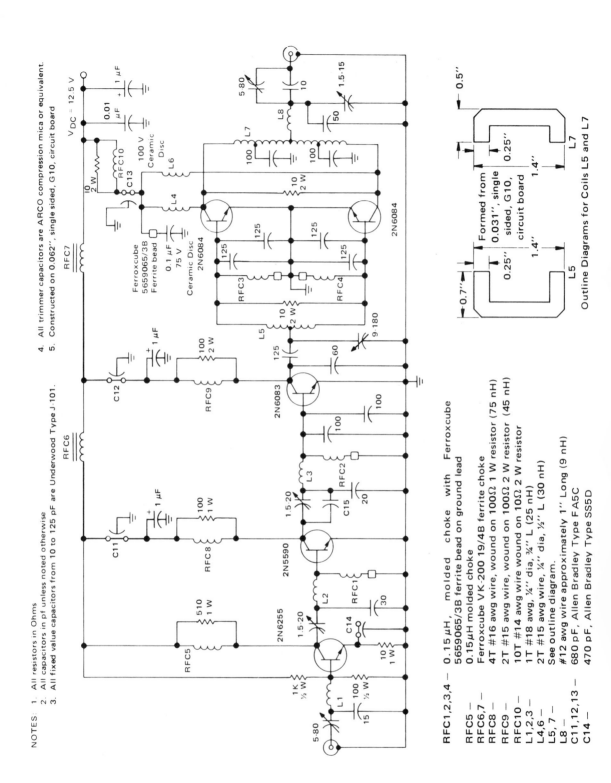

80-watt, 175-MHz Transmitter

This four-stage power amplifier is capable of producing 80 watts of continuous output power when operated from a 12.5-V supply. All the transmitter stages are of the common emitter configuration and, except for the input stage, all are operated Class C. The input stage is forward biased at approximately 40-mA collector current with no signal applied to improve performance at extremely low input signal levels. The high power output capability is achieved by operating two 2N6084 devices in parallel for the output stage. Source: John Hatchett, "Design Techniques for an 80-watt, 175 MHz Transmitter for 12.5 Volt Operation," Motorola Semiconductor Products Application Note AN-577, Copyright Motorola, Inc. Used by permission.

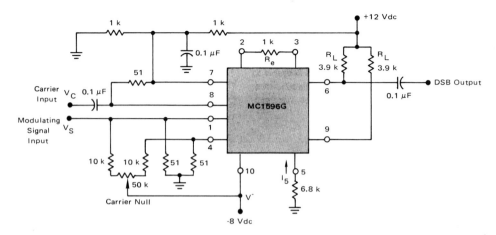

Balanced Modulator

This circuit shows the Motorola MC1596 in a balanced modulator circuit operating with +12- and −8-V supplies. You can obtain excellent gain and carrier suppression with this circuit by operating the upper (carrier) differential amplifiers at a saturated level and the lower differential amplifier in a linear mode. The recommended input signal levels are 60 mV RMS for the carrier and 300 mV RMS for the maximum modulating signal levels. Operating this circuit with a high level carrier input has the advantage of maximizing device gain and insuring that any amplitude variations present don't appear on the output sidebands. It has the disadvantage of increasing some of the spurious signals. Source: Roy Hejhall, "MC1596 Balanced Modulator." Motorola Semiconductor Products, Application Note AN-531, Copyright Motorola, Inc. Used by permission.

21. TRANSMITTING CIRCUITS

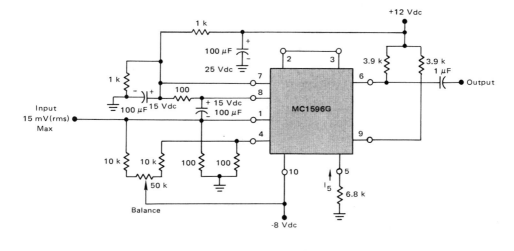

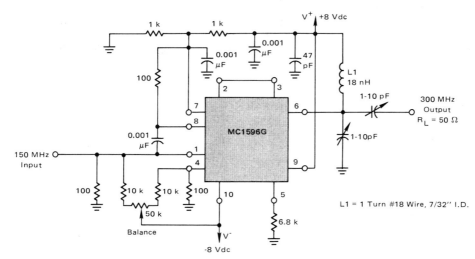

Frequency Doublers

The Motorola MC1596 balanced modulator functions as a frequency doubler when the same signal is injected in both input ports. The first circuit is a low frequency doubler that works below 1 MHz with all spurious outputs greater than 30 dB below the desired output signal. The second circuit is a 150- to 300-MHz doubler with output filtering. All spurious outputs are 20 dB below 300 MHz. Source: Roy Hejhall, "MC1596 Balanced Modulator." Motorola Semiconductor Products, Application Note AN-531, Copyright Motorola, Inc. Used by permission.

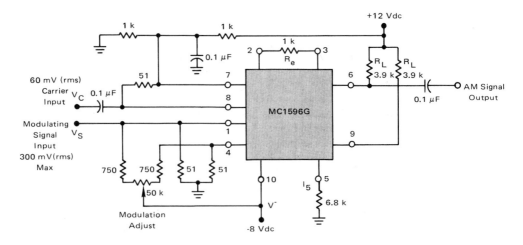

Amplitude Modulator

When connected as shown here, the Motorola MC1596 balanced modulator will function as an amplitude modulator. It will provide excellent modulation at any percentage from zero to greater than 100%. Source: Roy Hejhall, "MC1596 Balanced Modulator." Motorola Semiconductor Products, Application Note AN-531, Copyright Motorola, Inc. Used by permission.

21. TRANSMITTING CIRCUITS

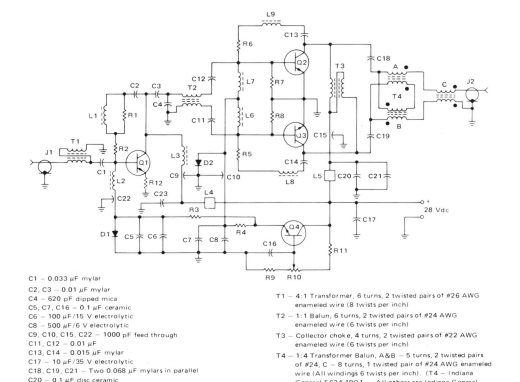

C1 — 0.033 µF mylar
C2, C3 — 0.01 µF mylar
C4 — 620 pF dipped mica
C5, C7, C16 — 0.1 µF ceramic
C6 — 100 µF/15 V electrolytic
C8 — 500 µF/6 V electrolytic
C9, C10, C15, C22 — 1000 pF feed through
C11, C12 — 0.01 µF
C13, C14 — 0.015 µF mylar
C17 — 10 µF/35 V electrolytic
C18, C19, C21 — Two 0.068 µF mylars in parallel
C20 — 0.1 µF disc ceramic
C23 — 0.1 µF disc ceramic
R1 — 220 Ω, 1/4 W carbon
R2 — 47 Ω, 1/2 W carbon
R3 — 820 Ω, 1 W wire W
R4 — 35 Ω, 5 W wire W
R5, R6 — Two 150 Ω, 1/2 W carbon in parallel
R7, R8 — 10 Ω, 1/2 W carbon
R9, R11 — 1 k, 1/2 W carbon
R10 — 1 k, 1/2 W potentiometer
R12 — 0.85 Ω (6 5.1 Ω or 4 3.3 Ω 1/4 W resistors in parallel, divided equally between both emitter leads)

T1 — 4:1 Transformer, 6 turns, 2 twisted pairs of #26 AWG enameled wire (8 twists per inch)
T2 — 1:1 Balun, 6 turns, 2 twisted pairs of #24 AWG enameled wire (6 twists per inch)
T3 — Collector choke, 4 turns, 2 twisted pairs of #22 AWG enameled wire (6 twists per inch)
T4 — 1:4 Transformer Balun, A&B — 5 turns, 2 twisted pairs of #24, C — 8 turns, 1 twisted pair of #24 AWG enameled wire (All windings 6 twists per inch). (T4 — Indiana General F624-19Q1, — All others are Indiana General F627-8Q1 ferrite toroids or equivalent.)

PARTS LIST

L1 — .33 µH, molded choke
L2, L6, L7 — 10 µH, molded choke
L3 — 1.8 µH (Ohmite 2-144)
L4, L5 — 3 ferrite beads each
L8, L9 — .22 µH, molded choke

Q1 — 2N6370
Q2, Q3 — **MRF460**
Q4 — 2N5190
D1 — 1N4001
D2 — 1N4997

J1, J2 — BNC connectors

Broadband 160-watt Linear Amplifier

This RF amplifier can supply 160 watts (PEP) into a 50-ohm load with IMD performance of -30 dB or better. Two Motorola MRF464 transistors are used in the design. For broadband linear operation, a quiescent collector current of 60–80 mA for each transistor should be provided. Higher quiescent current levels will reduce fifth-order IMD products, but will have little effect on third-order products except at low power levels. A biasing adjustment is provided in the amplifier circuit to compensate for variations in transistor current gain. This adjustment allows control of the idling current for both the output and driver devices. This control is also useful if the amplifier is operated from a supply other than 28 volts. Even with the biasing control, it's a good idea to beta match the output transistors. As with any push–pull design, both DC current gain and power gain at a midband frequency should be matched within about 15–20%. Source: Heige Granberg, "Broadband Linear Power Amplifiers using Push-Pull Transistors," Motorola Semiconductor Products, Application Note AN-593, Copyright Motorola, Inc. Used by permission.

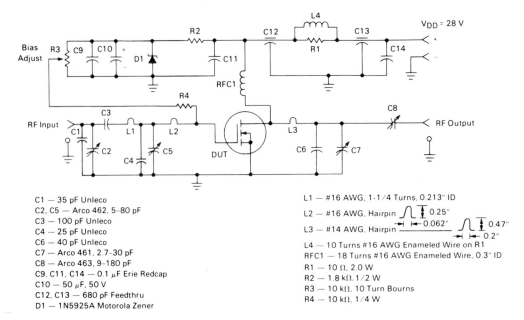

C1 — 35 pF Unleco
C2, C5 — Arco 462, 5-80 pF
C3 — 100 pF Unleco
C4 — 25 pF Unleco
C6 — 40 pF Unleco
C7 — Arco 461, 2.7-30 pF
C8 — Arco 463, 9-180 pF
C9, C11, C14 — 0.1 µF Erie Redcap
C10 — 50 µF, 50 V
C12, C13 — 680 pF Feedthru
D1 — 1N5925A Motorola Zener

L1 — #16 AWG, 1-1/4 Turns, 0.213" ID
L2 — #16 AWG, Hairpin 0.25"
L3 — #14 AWG, Hairpin 0.062" 0.47" 0.2"
L4 — 10 Turns #16 AWG Enameled Wire on R1
RFC1 — 18 Turns #16 AWG Enameled Wire, 0.3" ID
R1 — 10 Ω, 2.0 W
R2 — 1.8 kΩ, 1/2 W
R3 — 10 kΩ, 10 Turn Bourns
R4 — 10 kΩ, 1/4 W

Three VHF Amplifiers

These three VHF amplifiers are all designed around FETs. The first is a 150-watt, 150-MHz amplifier that operates from a 28 V DC supply. It has a typical gain of 12 dB, and can survive operation into a 30:1 VSWR load at any phase angle with no damage. The second is a simple 5-watt 150-MHz amplifier. The MRF134 used in this circuit is a very high gain FET which is potentially unstable at both VHF and UHF. A 68-ohm loading resistor has been used to enhance stability. This amplifier has a gain of 14 dB and a gain efficiency of 10.5 dB. The third circuit is a 5-watt, 400-MHz amplifier with a nominal gain of 10.5 dB. Source: Roy Hejhall, "VHF MOS Power Applications," Motorola Semiconductor Products, Application Note AN-878, Copyright Motorola, Inc. Used by permission.

21. TRANSMITTING CIRCUITS

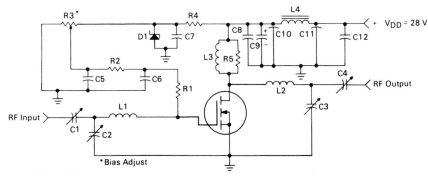

C1, C4 — Arco 406, 15-115 pF
C2 — Arco 403, 3-35 pF
C3 — Arco 402, 1.5-20 pF
C5, C6, C7, C8, C12 — 0.1 µF Erie Redcap
C9 — 10 µF, 50 V
C10, C11 — 680 pF Feedthru
D1 — 1N5925A Motorola Zener
L1 — 3 Turns, 0.310" ID, #18 AWG Enamel, 0.2" Long
L2 — 3-1/2 Turns, 0.310" ID, #18 AWG Enamel, 0.25" Long
L3 — 20 Turns, #20 AWG Enamel Wound on R5
L4 — Ferroxcube VK-200 — 19/4B
R1 — 68 Ω, 1.0 W Thin Film
R2 — 10 kΩ, 1/4 W
R3 — 10 Turns, 10 kΩ Beckman Instruments 8108
R4 — 1.8 kΩ, 1/2 W
R5 — 1.0 MΩ, 2.0 W Carbon
Board — G10, 62 mils

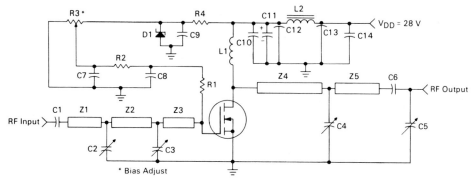

C1, C6 — 270 pF, ATC 100 mils
C2, C3, C4, C5 — 0-20 pF Johanson
C7, C9, C10, C14 — 0.1 µF Erie Redcap, 50 V
C8 — 0.001 µF
C11 — 10 µF, 50 V
C12, C13 — 680 pF Feedthru
D1 — 1N5925A Motorola Zener
L1 — 6 Turns, 1/4" ID, #20 AWG Enamel
L2 — Ferroxcube VK-200 — 19/4B
R1 — 68 Ω, 1.0 W Thin Film
R2 — 10 kΩ, 1/4 W
R3 — 10 Turns, 10 kΩ Beckman Instruments 8108
R4 — 1.8 kΩ, 1/2 W
Z1 — 1.4" × 0.166" Microstrip
Z2 — 1.1" × 0.166" Microstrip
Z3 — 0.95" × 0.166" Microstrip
Z4 — 2.2" × 0.166" Microstrip
Z5 — 0.85" × 0.166" Microstrip
Board — Glass Teflon, 62 mils

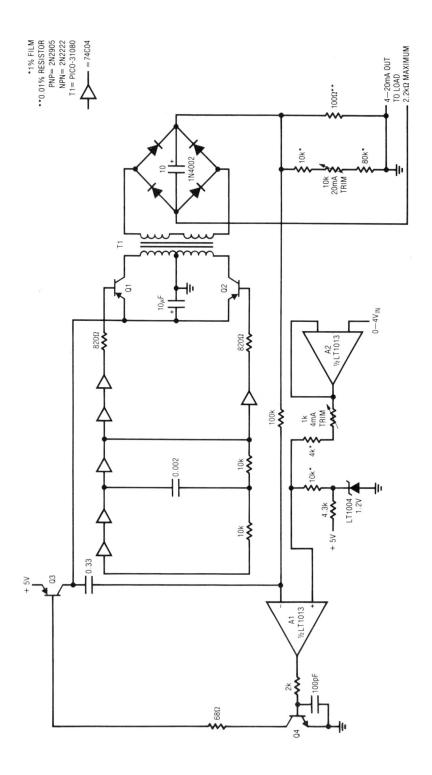

4- to 20-mA Current Loop Transmitter

Transmitting standard 4–20mA current loop signals to actuators is a common application. But resistive line losses and actuator impedances require current transmitters to to be able to force a compliance voltage of at least 20 V. Because of this, 5-V powered systems usually can't meet current loop transmitter requirements. But this circuit shows a way to do it. I- uses a servo-controlled DC–DC converter to generate the compliance voltage necessary for loop current requirements. To calibrate this circuit, short the output, apply 0 V to the input and adjust the "4-mA trim" for 0.3996 V across the 100-ohm resistor. Next, shift the input to 4.000 V and trim the 20-mA adjustment for 1.998 V across the 100-ohm resistor. Repeat this procedure until both points are fixed. Source: Jim Williams, "Designing Linear Circuits for 5V Operation," Application Note 11, Linear Technology Corporation.

Chapter 22
Video Circuits

Analog Multiplier Video Switch
Video Amplifier
Color Video Amplifier
Wide Bandwidth VCA for Video
Video Line Driving Amplifier
NTSC Video/Data Inlay Chip
IF Amplifier and Detector
Broadband Video Amplifier
IF Amplifier and Detector
NTSC Color Decoder
Wide-Band Video Amplifier
PAL/NTSC Decoder

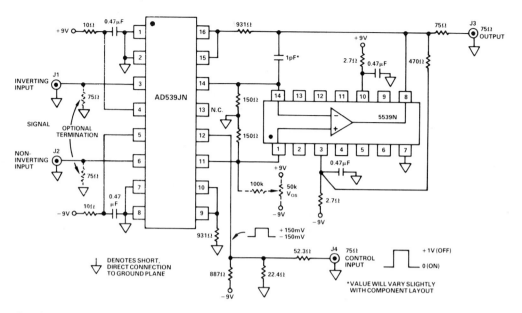

Analog Multiplier Video Switch

This fast video switch is suitable for many high-frequency applications including color-key switching. It features both inverting and noninverting inputs and can provide an output of ±1 V into a reverse-terminated 75-ohm load (or ±2 V into 150 ohms). This circuit provides a dimensionless gain of about one when on; zero when off. Source: Charles Kitchin, Andrew Wheeler, and Ken Weigel, "Low-Cost, Two-Chip Voltage-Controlled Amplifier and Video Switch," Application Note E957-20-11/85, Analog Devices.

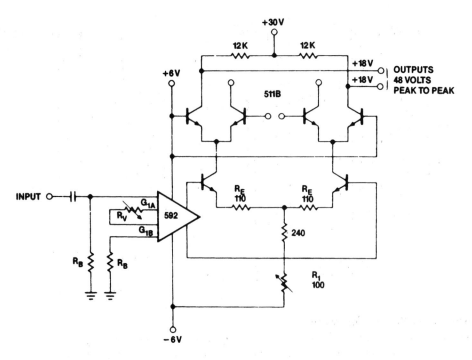

NOTE:
All resistor values are in ohms.

Video Amplifier

This circuit uses the Signetics NE/SE592 video amplifier. The NE592 drives an NE511B transistor array connected as a differential cascode amplifier. This circuit is capable of differential output voltages of 48 V p–p with a 3-dB bandwidth of approximately 10 MHz. For optimum operation, R1 is set for a no-signal level of +18 V. The emitter resistors (R_E) were selected to give the cascode amplifier a differential gain of 10. Source: "Using the NE/SE592 Video Amplifier," AN141, Linear Data Manual Volume 2: Industrial, Signetics Corporation.

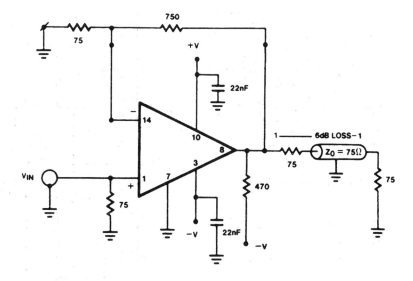

Color Video Amplifier

The Signetics NE5539 UHF op amp, as shown here, can be easily adapted for use as a color video amplifier. The gain of this circuit varies less than 0.5% across the video signal. The circuit is optimized for 75-ohm input and output impedance, and has a gain of about 10 (20dB). Source: Linear Data Manual Volume 2: Industrial, Signetics Corporation.

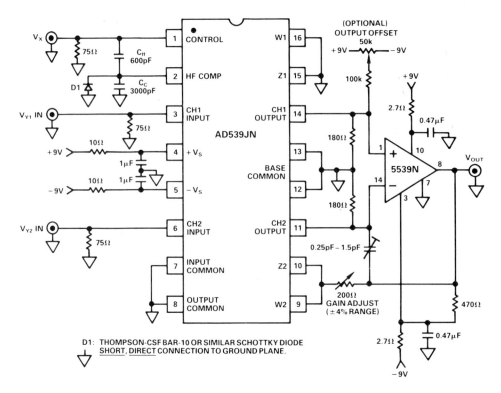

Wide Bandwidth VCA for Video

This circuit is a voltage-controlled amplifier (VCA) that's most suitable for video applications. Its bandwidth is over 50 MHz, and isn't substantially affected at lower gains. But when V_x is zero, there's a small amount of capacitance feedthrough at high frequencies. It's therefore important that you take extreme care in laying out the circuit board. Source: Charles Kitchin, Andrew Wheeler, and Ken Weigel, "Low-Cost, Two-Chip Voltage-Controlled Amplifier and Video Switch," Application Note E957-20-11/85, Analog Devices.

22. VIDEO CIRCUITS

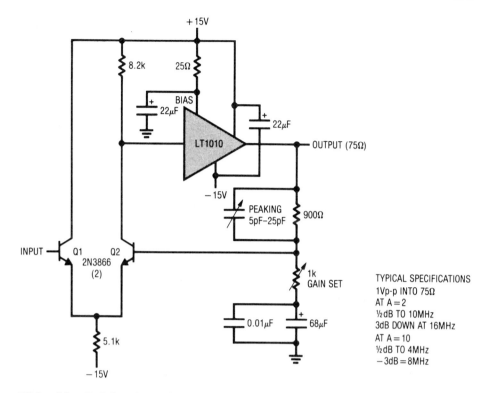

Video Line Driving Amplifier

In applications such as video, the DC stability of an amplifier is unimportant, and AC gain is required. This circuit combines the LT1010 power buffer's load handling capability with a fast, discrete gain stage. This circuit has a DC gain of 1, while allowing AC gains up to 10. Using a 20-ohm bias resistor, the circuit delivers 1 V p–p into a typical 75-ohm video load. If your application is sensitive to NTSC requirements, you can get better performance by dropping the value of the bias resistor. Source: Jim Williams, "Applications for a New Power Buffer," Application Note 4, Linear Technology Corporation.

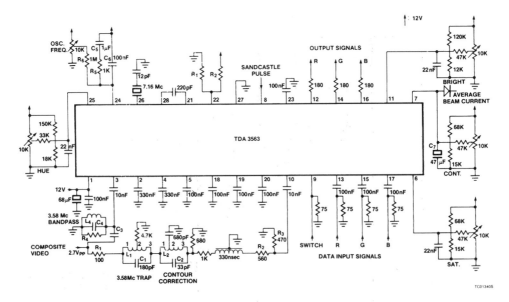

NTSC Video/Data Inlay Chip

The Signetics TDA3563 is a single-chip decoder that combines all the functions necessary for identification and demodulation of standard United States NTSC television signals. It also contains a luminance amplifier, an RGB matrix and RGB amplifiers, and analog inputs for external RGB data signals. Switching from video handling to data insertion occurs via fast video-data switching, that allows you to inlay data into a running picture. The circuit shown here is a typical application. Source: "Application of the NTSC Decoder: TDA3563," AN156, Linear Data Manual Volume 3: Video, Signetics Corporation.

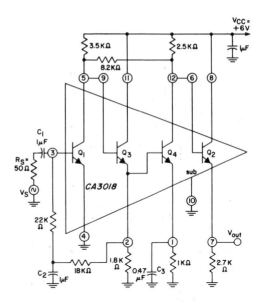

Broadband Video Amplifier

A common approach to video amplifier design is to use two transistors in a configuration designed to reduce the feedback capacitance. RCA's CA3018 IC, which has two isolated transistors plus two transistors with an emitter-base common connection is especially suited to an application like this, especially with its closely matched device characteristics. You can consider this amplifier as two DC-coupled stages, each consisting of a common-emitter, common-collector configuration. The common-collector transistor provides a low-impedance source to the input of the common-emitter transistor and a high-impedance, low-capacitance load at the common-emitter output. Two feedback loops provide DC stability and exchange gain for bandwidth. Source: G.E. Theriault, A.J. Leidich, and T.H. Campbell, "Application of the RCA-CA3018 Integrated-Circuit Transistor Array," Application Note ICAN-5296, RCA Electronic Components.

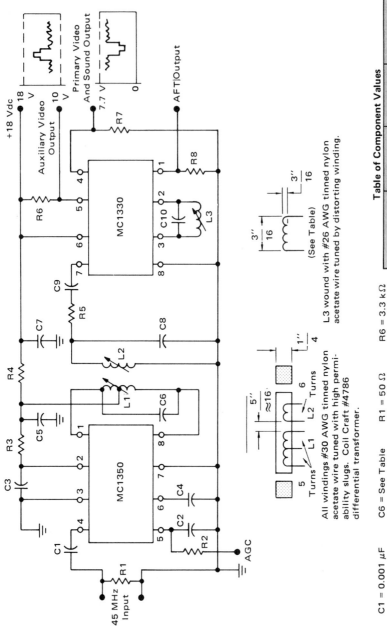

IF Amplifier and Detector

This circuit uses the Motorola MC1350 video amplifier and the MC1330 low level detector. It has a typical voltage gain of 84 dB and a typical AGC range of 80 dB. It also gives very small changes in bandpass shape, usually less than 1-dB tilt for 60-dB compression. Coupling between the two integrated circuits is achieved by a double-tuned transformer (L1 and L2). The sound intercarrier information may be taken from the detected video output. See the original application note for detailed information on aligning this circuit. Source: "Television Video IF Amplifier using Integrated Circuits," Motorola Semiconductor Products Application Note AN-545A, Copyright Motorola, Inc. Used by permission.

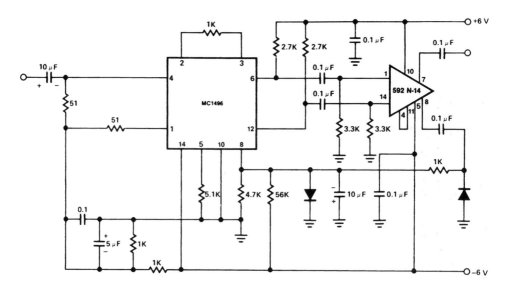

NOTE:
All resistor values are in ohms

Wide-Band AGC Video Amplifier

This circuit uses the Signetics NE592 along with an MC1496 balanced modulator to form an automatic gain control for video applications. The signal is fed to the signal input of the MC1496 and RC-coupled to the NE592. Unbalancing the carrier input of the MC1496 causes the signal to pass through unattenuated. Rectifying and filtering one of the NE592 outputs produces a DC signal which is proportional to the AC signal amplitude. After filtering, this control signal is applied to the MC1496 causing its gain to change. Source: "Using the NE/SE592 Video Amplifier," AN141, Linear Data Manual Volume 3: Video, Signetics Corporation.

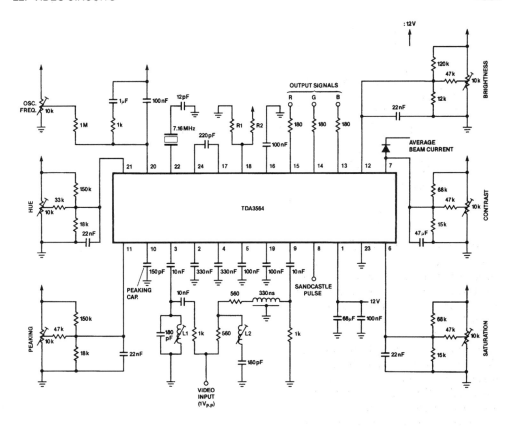

NTSC Color Decoder

The Signetics TDA3564 is a monolithic integrated decoder for the NTSC color television standard. It combines all the functions needed for the demodulation of NTSC signals. This circuit shows how it's used in a typical application. Source: Linear Data Manual Volume 3: Video, Signetics Corporation.

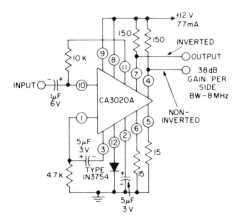

Wide-Band Video Amplifier

This design gives you an economical and stable class A bias for use in typical video applications. The differential portion of the CA3020A is placed at a potential above ground equal to the base-to-emitter voltage of the IC transistors (0.5 and 0.7 volts). This circuit provides a gain of 38 dB at each of the push–pull outputs, or 44 dB in a balanced-output connection. The 3-dB bandwidth of the circuit is 30 Hz to 8 MHz. You can get higher gain-bandwidth performance if the diode-to-ground voltage drop at terminal 12 is reduced. Source: W. M. Austin and H. M. Kleinman, "Application of the RCA CA3020 and CA3020A Integrated-Circuit Multi-Purpose Wide-Band Power Amplifiers," Application Note ICAN-5766, RCA Solid State.

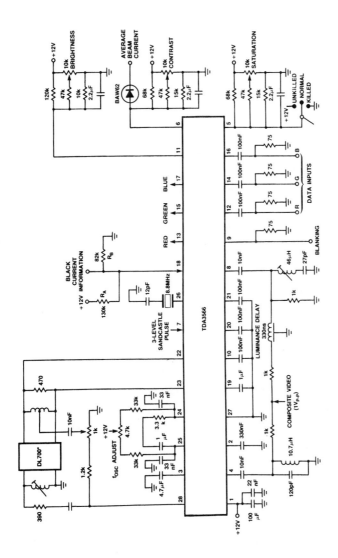

NOTE:
*D1700 AMPEREX CORP

PAL/NTSC Decoder

If you have a requirement for a television application that requires access to both United States NTSC and European PAL standards, the Signetics TDA3566 PAL/NTSC Decoder with RGB inputs has all circuits required for the demodulation of both. It also has data display capabilities. This circuit is a typical application.
Source: Linear Data Manual Volume 3: Video, Signetics Corporation.

Chapter 23

Voltage Circuits

Fast, Synchronous Rectifier-Based AC/DC Converter
Visible Voltage Indicator
12-nsec Circuit Breaker
Offset Stabilized Comparator
−5-volt Bus Monitor
+6- to +15-volt Converter
Regulated Voltage Up Converter
Precision Voltage Inverter
50-MHz Bandwidth RMS-to-DC Converter
Standard-Grade Variable Voltage Reference
Dual-Voltage Tracking Regulator
Isolated Power Line Monitor
Dual-Limit Threshold Detector
LVDT Signal Conditioner

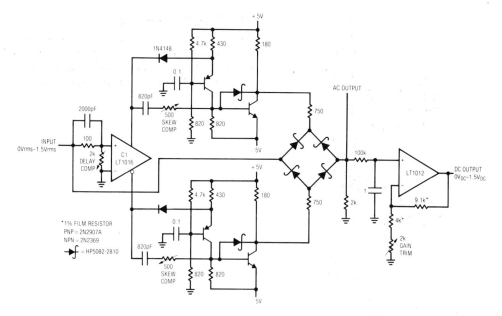

Fast, Synchronous Rectifier-Based AC/DC Converter

Most precision rectifier circuits rely on operational amplifiers to correct for diode drops. Although this works well, bandwidth limitations usually restrict these circuits to operation below 100 KHz. But this circuit uses a LT1016 comparator in an open-loop, synchronous rectifier configuration for high accuracy out to 2.5 MHz. To calibrate this circuit, apply a 1- to 2-MHz 1-V p–p sine wave and adjust the delay compensation so bridge switching occurs when the sine crosses zero. Next, adjust the skew compensation potentiometers for minimum aberrations in the AC output signal. Source: Jim Williams, "High Speed Comparator Techniques." Application Note 13, Linear Technology Corporation.

Visible Voltage Indicator

This very simple yet very effective visible voltage indicator uses four LM139 voltage comparators and a handful of parts. Source: Linear Data Manual Volume 2: Industrial, Signetics Corporation.

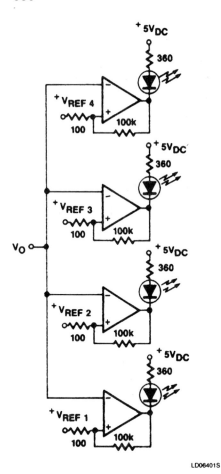

12-nsec Circuit Breaker

This simple circuit turns off current in a load just 12 ns after it exceeds a preset value. It's valuable for protecting integrated circuits during developmental probing, as well as for protecting expensive loads during trimming and calibration. Once the circuit is triggered, the LT1016 is held in a latched state by feedback from the noninverting output. When the load fault has been cleared, you can use the pushbutton to reset the circuit. Source: Jim Williams, "High Speed Comparator Techniques." Application Note 13, Linear Technology Corporation.

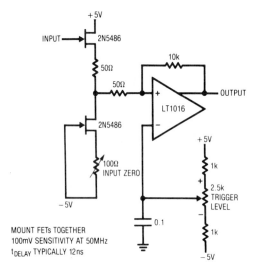

23. VOLTAGE CIRCUITS

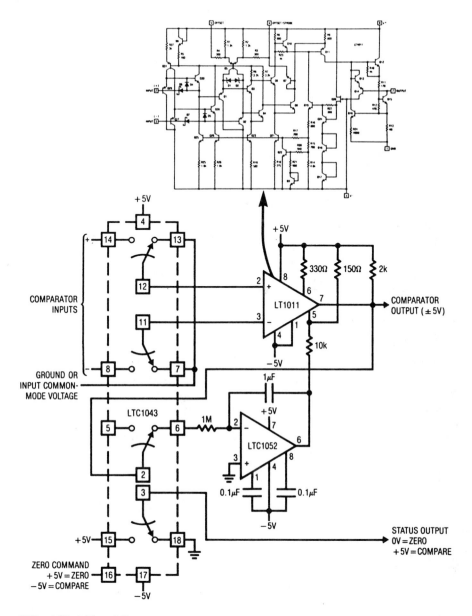

Offset Stabilized Comparator

A reasonably fast voltage comparator with low input offset drift is useful in a variety of applications such as high-resolution analog-to-digital converters, crossing detectors, and anywhere else where you need to make a precise, stable, and high-speed comparison. This circuit uses a Linear Technology LTC1052 chopper-stabilized amplifier to eliminate offset and drift without sacrificing speed or differential input versatility. Realize though, that this circuit is applicable only in situations where some dead time is available so that zeroing action can occur. Source: Jim Williams, "Application Considerations and Circuits for a New Chopper-Stabilized Op Amp," Application Note 9, Linear Technology Corporation.

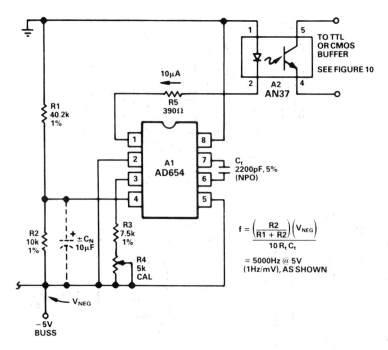

−5-volt Bus Monitor

This simple circuit is scaled to monitor a common −5-V bus voltage, with a 1.0-Hz/mV sensitivity. While this example is for a 5-V bus, a simple adjustment of the R1/R2 divider will allow the circuit to monitor other voltages. Source: Walt Jung, "Operation and Applications of the AD654 IC V-to-F Converter," Application Note E923-25-7/85, Analog Devices.

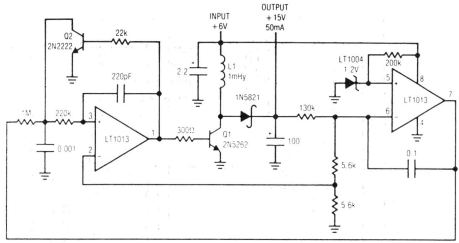

L1 = AIE—VERNITRON 24-104
78% EFFICIENCY

+6- to +15-volt Converter

With a relatively high voltage and high power output, this circuit is ideally suited to the requirements of mixed linear-digital circuits. From a 6-V battery, it delivers 15 V at up to 50 mA. Its regulation is within 0.05% over a wide range of output loads. Source: Jim Williams, "Power Conditioning Techniques for Batteries," Application Note 8, Linear Technology Corporation.

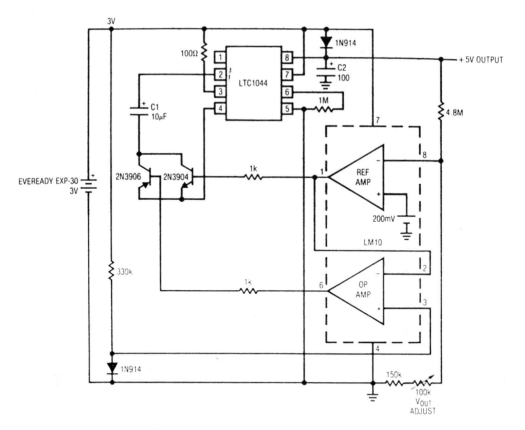

Regulated Voltage Up Converter

This circuit addresses the regulation fall-off of a simple voltage up converter by increasing current. It takes a 3-V input and fixes the 5-V output within 0.025 V for loads up to 2 mA. Source: Jim Williams, "Power Conditioning Techniques for Batteries," Application Note 8, Linear Technology Corporation.

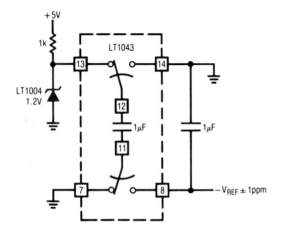

Precision Voltage Inverter

Featuring a high input impedance and requiring no trimming, this circuit lets you invert a reference voltage with an accuracy of 1 ppm. Source: Jim Williams, "Applications for a Switched-Capacitor Instrumentation Building Block," Application Note 3, Linear Technology Corporation.

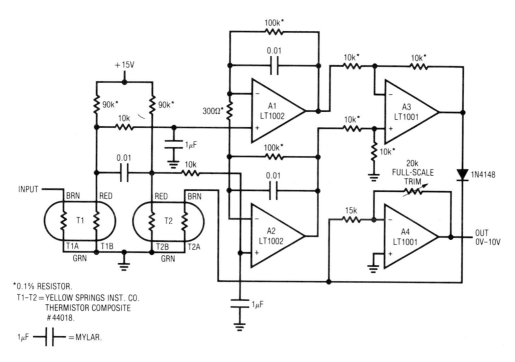

50-MHz Bandwidth RMS-to-DC Converter

Conversion of AC waveforms to their equivalent DC power value is usually accomplished by either rectifying and averaging or using analog computing methods. Rectification averaging works only for sinusoidal inputs, and analog computing methods are limited to use below 500 kHz. A way to achieve wide bandwidth and high crest factor performance is to measure the true power value of the waveform directly, which this circuit does by measuring the DC heating power of the input waveform. The original application note includes details on the recommended thermal arrangement for the thermistors using 2-inch spacing on a styrofoam block. To calibrate this circuit, apply 10 V DC to the input and adjust the full-scale trim for 10 volts out at A4. Source: Jim Williams, "Thermal Techniques in Measurement and Control Circuitry," Application Note 5, Linear Technology Corporation.

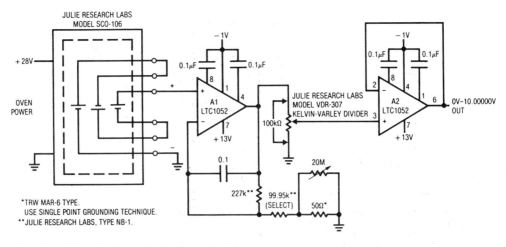

Standard-Grade Variable Voltage Reference

This laboratory grade variable voltage reference may be used to calibrate 6 1/2 digit voltmeters, ultra-high resolution data converters and other apparatus that require high order traceability to primary standards. To calibrate this circuit, adjust A1's output for exactly 10 V by selecting the feedback resistor and fine trimming the 20-Mohm potentiometer. Once calibrated, this circuit will provide worst-case 0.0014% accuracy for one year's time. Source: Jim Williams, "Application Considerations and Circuits for a New Chopper-Stabilized Op Amp," Application Note 9, Linear Technology Corporation.

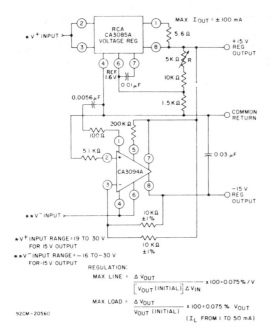

Dual-Voltage Tracking Regulator

In this circuit, the magnitude of the regulated positive voltage provided by the RCA CA3058A regulator is adjusted by potentiometer R. A sample of this positive regulated voltage supplies the power for the RCA CA3094 monolithic programmable power switch/amplifier, used here as the negative regulator, and also supplies a reference voltage to its terminal 3 to provide tracking. The regulator provides a maximum line regulation equal to 0.075 per volt of input voltage change and a maximum load regulation of 0.075% of the output voltage. Source: L. R. Campbell and H. A. Wittlinger, "Some Applications of a Programmable Power Switch/Amplifier." Application Note ICAN-6048, RCA Solid State.

Isolated Power Line Monitor

This circuit uses the Burr-Brown ISO102 signal isolation buffer amplifier to allow you to monitor power-line voltage while staying isolated from it. The differential input accurately senses the power resistor voltage, and two resistors are used to protect the IC from an open power resistor condition. This circuit has 0.5-μA leakage current at 120 V RMS. Source: "Integrated Circuits Data Book Supplement," Burr-Brown.

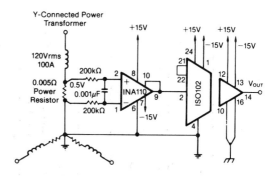

Differential input accurately senses power resistor voltage.
Two resistors protect INA110 from open power resistor.
High frequency spike reject filter has f_{co} = 400Hz.

Dual-Limit Threshold Detector

Also known as a comparator, this circuit uses potentiometer R1 to set the high-level limit, and potentiometer R2 for the low-level limit. R2 also actuates the RCA CA3094 monolithic programmable power switch/amplifier. A positive output signal is delivered by the CA3094 whenever the input signal exceeds either the high-limit or low-limit values established by the potentiometers. As shown, the output voltage is approximately 12 volts. Source: L. R. Campbell and H. A. Wittlinger, "Some Applications of a Programmable Power Switch/Amplifier." Application Note ICAN-6048, RCA Solid State.

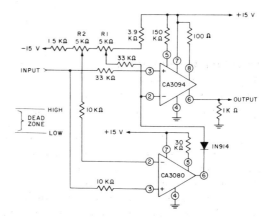

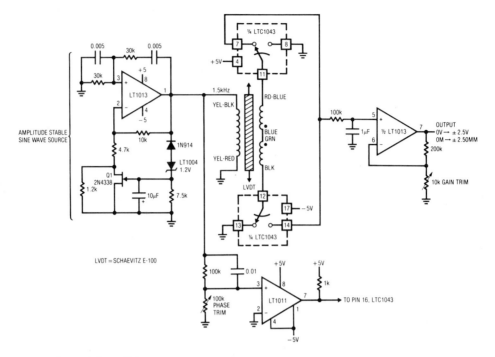

LVDT Signal Conditioner

A linear variable differential transformer (LVDT) is a transformer with a mechanically actuated core. The primary is driven by a sine wave, usually amplitude stabilized. The sine drive eliminates error-inducing harmonics. The two secondaries are connected in opposed phase. When the core is positioned in the magnetic center of the transformer, the secondary outputs cancel and there's no output. Moving the core away from the center position unbalances the flux ratio between the secondaries, developing an output. To calibrate this circuit, center the LVDT core in the transformer and adjust the phase trim for 0-V output. Next, move the core to either extreme position and set the gain trim for 2.50-V output. Source: Jim Williams, "Applications for a Switched-Capacitor Instrumentation Building Block," Application Note 3, Linear Technology Corporation.

Index

Amplifiers:
   +/− 5V precision instrumentation, 21
   AC-coupled inverting, 27
   chopper-stabilized instrumentation, 20
   current mode feedback, 28
   DC-coupled inverting, 18
   DC-coupled, non-inverting, 19
   DC-stabilized low noise, 26
   differential input/differential output, 36
   fast DC stabilized FET, 29
   fast-stabilized non-inverting, 31
   FET, 29, 32
   gain trimmable wideband FET, 32
   instrumentation, 20, 21, 23
   inverting, 18, 27,
   isolation, 24–25
   large signal-swing output, 23
   lock-in, 17
   low-noise, 26
   low-power voltage boosted output operational, 33
   N-stage parallel-input, 35
   operational, 22, 33
   parallel input, 35
   precision high-speed operational, 22
   precision isolation amplifier, 24–25
   precision with notch, 36
   single-supply differential bridge, 34
   three-channel separate-gain, 34
   ultra-precision instrumentation, 23
   variable gain, 16
   voltage-controlled, 30
Amplifiers, audio:
   100-watt, 12–13
   12-watt, 5
   15-, 20-, and 25-watt Darlington, 6
   20-watt, 7
   3- to 5-watt, 3–4
   310-milliwatt transformerless, 2
   6-watt, automotive, 61
   7- to 35-watt, 8–9
   basic class B, 2
   ceramic pickup, 13
   class B, 2
   Darlington, 6
   gain-controlled stereo, 13
   high power, 10–11
   phono pickup, 13
   stereo, 13
   transformerless, 2
Audio Circuits:
   4-Input stereo source selector, 45
   AGC amplifier, 49
   amplifier, AGC, 49
   attenuator, voltage controlled, 40
   audio decibel level detector, 47
   audio generator, 49
   automatic level control, 41
   compandor, 43–44
   compressor-expander, 42
   detector, level, 47
   equalizer, 50–51
   fast attack, slow release hard limiter, 47
   filter, rumble/scratch, 50
   generator, audio, 49
   hi-fi compandor, 43–44
   level detector, 47
   limiter, 47
   microphone preamplifier, 38
   music synthesizer, 39
   preamplifier, microphone, 38
   preamplifier, RIAA/NAB, 44, 46, 53
   preamplifier, tape head 41,
   RIAA equalized stereo preamplifier, 46
   RIAA preamplifier, 53
   RIAA/NAB compensation preamplifier, 44
   rumble/scratch filter, 52
   single-frequency audio generator, 49
   stereo source selector, 45
   stereo volume control, 54
   synthesizer, music, 39
   tape head preamplifier, 41
   variable-slope compressor-expander, 42
   voltage-controlled attenuator, 40
   volume control, stereo, 54

INDEX

Automotive Circuits:
  6-watt audio amplifier, 61
  AM radio, 61
  AM/FM radio, 60
  burglar alarm, 57
  speed warning device, 58
  tachometer, 56
  voltage regulator, 59

Battery Circuits:
  6-volt charger, 64–65
  12-volt charger, 64
  backup regulator, 66
  cell monitor, 67
  charger, 12-volt, 64
  charger, 6-volt, 64–65
  charger, NiCad, 71
  charger, wind-powered, 65
  charging circuit monitor, 71
  computer battery control, 74–75
  converter, sine wave output, 68
  doubler, voltage, 70
  high-current splitter, 67
  low dropout 5-volt regulator, 73
  low-power flyback regulator, 70
  micropower switching regulator, 68
  monitor for high-voltage charging circuits, 71
  monitor, battery cell, 67
  regulator, 5-volt, 73
  regulator, backup, 66
  regulator, flyback, low-power, 70
  regulator, switching, micropower, 68
  sine wave output converter, 69
  single-cell up converter, 72
  splitter, high-current, 67
  thermally controlled NiCad charger, 71
  up converter, single-cell, 72
  voltage doubler, 70
  wind-powered charger, 65

Control Circuits:
  3-phase sine wave circuit, 87
  alarm system, 83
  automatic light control, 93
  controller, motor speed control, 83
  controller, temperature, 82, 84, 85, 87, 88, 89
  controller, zero-voltage, 77
  decoder, touch tone, 81, 86, 92
  dual time-constant tone decoder, 86
  dual-output over-under temperature controller, 87
  fan-based temperature controller, 84
  frequency meter with lamp readout, 91
  line-isolated temperature controller, 85
  meter, frequency, 91
  motor speed control, 93
  overload protected motor speed controller, 83
  precision temperature controller, 82
  sensitive temperature controller, 88
  squelch control, 90
  switch, touch, 81
  tachless motor speed controller, 78
  temperature control, 89
  thermostat, wall-type, 79
  touch switch, on/off, 81
  touch-tone decoder, low-cost, 81
  wall-type thermostat, 79
  wide-band tone detector, 92
  zero-voltage, on-off controller w/sensor, 77

Converters:
  1-Hz to 1.25-MHz voltage-to-frequency, 95
  1-Hz to 30-MHz voltage-to-frequency, 96
  10-bit 100-μA analog-to-digital, 104
  16-bit analog-to-digital, 97
  2.5-MHz fast-response V/F, 108
  9-bit digital-to-analog, 102
  analog-to-digital, 97, 98, 104, 105
  basic 12-bit 12-μsec successive approximation, 114
  centigrade-to-frequency, 116–117
  current loop receiver/transmitter, 100–101
  current-to-voltage, 98
  cyclic analog-to-digital, 98
  digital-to-analog, 102,
  ECL-to-TTL, 112
  frequency-to-voltage, 115
  fully-isolated 10-bit A/D, 106
  Kelvin-to-frequency, 116–117
  low-power 10-bit A/D, 113
  low-power 10-KHz voltage-to-frequency, 111
  micropower 1-MHz voltage-to-frequency, 119
  micropower 10-kHz voltage-to-frequency, 118
  micropower 12-bit 300-μsec analog-to-digital, 105
  offset stabilized voltage-to-frequency, 99
  push-pull transformer-coupled, 109
  quartz-stabilized voltage-to-frequency, 107
  RMS-to-DC, 120
  successive approximation, 114

temperature-to-frequency, 103
thermocouple-to-frequency, 101
transformer coupled, 109
TTL-to-ECL, 112
TTL-to-MOS, 102
ultra-high-speed 1-Hz to 100-MHz V/F, 110
V/F-F/V data transmission circuit, 106–107
voltage-to-frequency, 95, 96, 99, 107, 111, 118, 119
wide-range precision PLL frequency-to-voltage, 115

Filter Circuits:
  10-Hz, fourth-order Butterworth low-pass, 127
  60-Hz reject, 131
  bi-quad, 128
  Butterworth, 127
  clock tunable notch, 127
  digitally-tuned switched capacitor, 126
  loop, 122
  low-pass, 122
  lowpass with 60-Hz notch, 131
  multi-cutoff-frequency, 130
  notch, 127, 131
  rumble/scratch, 50
  second-order, 123
  switched capacitor, 126
  tracking, 125
  tunable, 127
  voltage-controlled, 124
  voltage-controlled, 129
  voltage-controlled, 132
  voltage-controlled, second-order, 123

Function Generators:
  function generators, 134, 140
  logic function, 135
  single-chip, 137
  triangle wave oscillator, 136
  triangle-square wave, 136
  triangle-to-sine wave, 138
  waveform generator, 139

Measurement Circuits:
  4-channel temperature sensor, 161
  accelerometer, 143
  acoustic thermometer, 148–149
  anemometer, 155
  digital thermometer, 157, 165
  field strength meter, 164
  flowmeter, thermal, 153
  level transducer digitizer, 159
  linear thermometer, 142, 151
  long-duration timer, 165
  low flow rate thermal flowmeter, 153
  magnetic tachometer, 150
  meter, field strength, 164
  monitor, AC current, 158
  monitor, current, 147
  narrow-band tone detector, 146
  negative current monitor, 147
  photodiode digitizer, 162–163
  presettable timer, 154
  relative humidity signal conditioners, 157
  sequential timer, 144
  sine-wave averaging AC current monitor, 158
  tachometer, 164, 150
  tachometer, automotive, 56
  thermally based anemometer, 155
  thermocouple amplifier, 145
  thermometer, 142, 148–149, 150, 151, 157, 160
  timer, 144, 154
  transmitting thermometer, 150
  voltmeter, 152

Microprocessor Circuits:
  8-bit serial-to-parallel converter, 167
  bidirectional bus interface, 171
  Cheapernet/Ethernet interface, 172
  clock regenerator, 170
  converter, serial-to-parallel, 167
  CRT driver, 169
  detector, voltage sag, 167
  driver, CRT, 169
  interface, bus, 171
  interface, LAN, 172
  power-loss detection circuit, 168
  voltage-sag detector, 167

Miscellaneous Circuits:
  alarm, freezer, 180
  analog divider, 183
  astable single-supply multivibrator, 177
  buffer, 178
  buffered output line driver, 175
  clock sources, 179
  digital clock with alarm, 176
  divider, analog, 183
  fast, precision sample-hold, 186
  fed forward, wideband DC stabilized buffer, 178

Miscellaneous Circuits (*continued*)
   fiber optic receiver, 181
   freezer alarm, 180
   frequency output analog divider, 183
   frequency synthesizer, 177
   intercom, 174
   lamp driver, 183
   line driver, buffered output, 175
   low-voltage lamp flasher, 174–175
   micropower sample-hold, 182
   multivibrator, 177
   peak circuits, 188
   PLL, precision, 187
   precision PLL, 187
   protected high current lamp driver, 183
   receiver, fiber optic, 181
   sample-hold, 182
   shift register, 185
   synthesizer, 177
   tone transceiver, 184
   transceiver, tone, 184
   variable shift register, 185
Modem Circuits:
   1200 bps, 193
   2400 bps CCITT, 197
   2400 bps stand-alone intelligent, 199
   300 bps full-duplex, 192
   300/1200 bps, 194–195
   auto dialer circuit, 198
   Bell 212A, 190
   FSK, 201, 202–203
   full-duplex 300/1200 bps, 194–195
   full-duplex FSK, 201
   high-speed FSK, 202–203
   notch filters, 200
   power line, 204

Optoelectronics Circuits:
   100-dB range logarithmic photodiode amplifier, 212–213
   amplifier, light, 207
   amplifier, photodiode, 209, 212–213
   balanced pyroelectric infrared detector, 211
   detector, infrared, 211
   detector, photodiode, 210
   driver, LED, 206
   high-speed photodetector, 210
   infrared remote-control system, 208–209
   LED driver, 206
   light amplifier, 207
   photodiode amplifiers, 207, 209
   photodiode detector, 210
   remote-control, infrared, 208–209
   sensitive photodiode amplifier, 209
Oscillator Circuits:
   1- to 10-MHz crystal, 215
   1- to 25-MHz crystal, 215
   1-Hz to 1-MHz sine wave output voltage controlled, 224
   1st harmonic (fundamental), 227
   crystal, 215, 218, 221, 222
   crystal-controlled, 221
   crystal-stabilized relaxation, 216
   digitally-programmable PLL, 226
   high-current, 221
   L-C tuned, 217
   low distortion sinewave, 228
   low frequency precision RC, 227
   low-power temperature compensated, 225
   PLL, 226
   relaxation, 216
   reset stabilized, 220
   sinewave, 228
   stable RC, 219
   synchronized, 221
   temperature compensated, 225
   temperature-compensated crystal, 218, 230
   voltage controlled, 223, 224
   voltage-controlled crystal, 222
   wein bridge, 229

Power Supply Circuits:
   5-V regulator with shutdown, 239
   5-V regulator, 239
   7.5-A variable regulator, 255
   adjustable regulator, 256
   basic, 232
   bridge amplifier load current monitor, 237
   current monitor, 235
   current regulator, 251
   current-limited 1-amp regulator, 238
   dual output regulator, 240
   dual tracking voltage regulator, 250
   fully-isolated −48 V to 5-V regulator, 247
   high current switching regulator, 244
   high-efficiency rectifier circuit, 238
   high-efficiency regulator, 257
   high-temperature +15-V regulator, 255
   high-voltage regulator, 253

inductorless switching regulator, 248
low-power switching regulator, 246
micropower post-regulated switching regulator, 242–243
negative-voltage regulator, 252
programmable voltage/current source, 234
rectifiers, 236
regulated negative voltage converter, 235
regulator, automotive, 59
regulator with output voltage monitor, 241
RMS voltage regulator, 242
single inductor, dual-polarity regulator, 249
switching preregulated linear regulator, 245
switching regulator, 243, 254
TTL monitor, 233

Preamplifiers
microphone preamplifier, 38
preamplifier, RIAA/NAB, 44, 46, 53
preamplifier, tape head 41

Receiving Circuits:
10.7-Hz IF amplifier, 265
10.8-MHz FSK decoder, 263
4–20 mA current receiver, 267
88- to 108-MHz FM front end, 266
amplifier, IF, 265
AM radio, automotive, 61
AM/FM radio, automotive, 60
balanced mixer, 267
clock regenerator, 262
decoder, FSK, 263
demodulator, FM, 261, 264
demodulator, SCA, 259
detector, FM, 264
FM detector, 264
FM tuner, 260
IF amplifier, 265
linear FM detector, 264
narrow-band FM demodulator, 264
narrow-bandwidth FM demodulator, 261
SCA demodulator, 259

Regulator Circuits:
5-V with shutdown, 239
5-V, 239
7.5-A variable, 255
adjustable, 256
automotive, 59
backup, 66
current, 251
current-limited 1-amp, 238
dual output, 240
dual tracking voltage, 250
fully-isolated −48 V to 5 V, 247
high current switching, 244
high-efficiency, 257
high-temperature +15-V, 255
high-voltage, 253
inductorless switching, 248
low-dropout, 5-volt, battery, 73
low-power flyback, battery, 70
low-power switching, 246
micropower post-regulated switching, 242–243
negative-voltage, 252
RMS voltage, 242
single inductor, dual-polarity, 249
switching preregulated linear, 245
switching, 243
switching, 254
with output voltage monitor, 241

Signal Circuits:
2.2-Watt incandescent lamp driver, 270
automatic gain control, 271
bandpass filter for tone detector, 274–275
high-speed peak detector, 272
lamp driver, 270
low-droop positive peak detector, 271
peak detector, 272
single-burst tone generator, 273
tone burst generator, 270
tone generator, 273, 274–275
traffic flasher, 269
wide-range automatic gain control, 271

Telephone Circuits:
audio frequency sweepers, 284–285
basic telephone set, 279
bell system 202 data encoder/decoder, 286
featurephone with memory, 282–283
line-powered tone ringer, 277
non-isolated 48V to 5V regulator, 280
programmable multi-tone telephone ringer, 281
regulator, telephone voltage, 281
ring detector, 288
ring signal counter, 287
ringer, tone, 277
tone ringer, 277
tone telephone, 278

Test Circuits:
   0–5 Amp, 7–30 Volt laboratory supply, 291
   1ns rise time pulse generator, 299
   50-MHz trigger, 299
   active and passive signal combiners, 300
   amplifier, gain-ranging, 308
   amplifier, instrumentation, 307
   analog multiplier, 294
   cable tester, 303
   combiners, signal, 300
   computer-controlled digitizer, 296–297
   digitally-programmable frequency source, 305
   digitally-programmable waveform generator, 298
   digitizer, 296–297
   gain-ranging amplifier, 308
   generator, pulse, 292, 299, 304
   instrumentation amplifier, 307
   leakage current monitor, 295
   low-frequency pulse generator, 304
   low-noise instrumentation amplifier, 307
   precision programmable voltage reference, 20
   pulse generator, 292
   signal combiners, 300
   synthesizer, 293
   tester, cable, 303
   ultraprecision variable voltage reference, 307
   voltage programmable pulse generator, 306
   voltage reference, 20, 298, 301, 307
   waveform generator, 298
   wide-range frequency synthesizer, 293

Transmitting Circuits:
   4–20 mA current loop transmitter, 318–319
   80-watt, 175-MHz transmitter, 310–311
   amplitude modulator, 314
   balanced modulator, 312
   broadband 160-watt linear amplifier, 315
   frequency doublers, 313
   VHF amplifiers, 316–317

Video Circuits:
   amplifier, 322
   analog multiplier video switch, 321
   broadband amplifier, 327
   color amplifier, 323
   IF amplifier and detector, 328–329
   NTSC color decoder, 331
   NTSC video/data inlay chip, 326
   PAL/NTSC decoder, 333
   video line driving amplifier, 325
   wide bandwidth VCA, 324
   wide-band AGC amplifier, 330
   wide-band amplifier, 332